Lecture Notes in Economics and Mathematical Systems

Antoon van den Elzen

Adjustment Processes for Exchange Economies and Noncooperative Games

Springer-Verlag
Berlin Heidelberg New York
London Paris Tokyo
Hong Kong Barcelona
Budapest

Author

Dr. Antoon van den Elzen
Tilburg University
Department of Econometrics
P. O. Box 90153
NL-5000 LE Tilburg, The Netherlands

ISBN 978-3-540-57310-4 ISBN 978-3-642-48133-8 (eBook)
DOI 10.1007/978-3-642-48133-8

Typesetting: Camera ready by authors
42/3140-543210 - Printed on acid-free paper

ACKNOWLEDGEMENTS

This monograph is the result of research carried out over a period of seven years. The first two years the research was financially supported by Tilburg University. The other years by the cooperation center of Eindhoven University and Tilburg University.

This work would never have been completed without the help of several people. First of all, I mention Dolf Talman who introduced me to the field of simplicial algorithms and related topics. His enthusiasm and guidance encouraged and stimulated me very much. Next, I would like to thank Pieter Ruys who always supported me, also on times when things went not so well. Further, I am indebted to Gerard van der Laan with whom I did part of this research. Jan van Geldrop introduced me to the field of differential topology, and Tim Doup made me overcome my initial fear for computers.

I thank Petra Ligtenberg for her excellent typewriting, and I am indebted to Yvonne van Delft and Jan Pijnenburg for drawing the nice figures. More generally, I would like to thank all my roommates and the colleagues at the Department of Econometrics of Tilburg University for the excellent environment they created for doing scientific research.

Antoon van den Elzen

Contents

CHAPTER 1

Introduction

This monograph is a treatise on adjustment processes. We consider price adjustment processes in exchange economies and strategy adjustment processes in noncooperative games.

In the most simple version of an exchange economy, i.e. a pure exchange economy, there exist markets on which prices are determined by the demand and supply created by a finite number of consumers willing to exchange their initial endowments in order to maximize their utilities. An equilibrium situation is attained if, for some price vector, demand equals supply in all markets. Starting from a situation not being an equilibrium an adjustment process reaches an equilibrium via adaptations of prices. The advantage of the adjustment processes we will present in this monograph is that they exist and converge under far weaker assumptions than existing processes.

The second subject concerns the problem of finding Nash equilibria in noncooperative games. A Nash equilibrium is a situation from which no player can improve his position by unilaterally changing his strategy. We present a new algorithm for finding such equilibria. The sequence of strategy vectors generated by the algorithm can be interpreted as the path followed by a strategy adjustment process.

This introductory chapter consists of three sections. In Section 1.1 we give an overview of the existing literature on price adjustment processes in exchange economies, discuss the meaning and value of these processes, and consider the contribution of this monograph against this background. Next, in Section 1.2 we illustrate both our type of price adjustment processes and the algorithm for finding Nash equilibria, by giving two simple examples. Finally, Section 1.3 gives an outline of the monograph and sketches the main results.

1.1. Adjustment processes in economies

In this section we want to introduce the concept of an adjustment process and discuss its significance. We do this by considering the main processes that have been defined in the literature. Also, we indicate our contribution to the field. For a comprehensive overview of the literature we refer to the articles of Hahn [1982] and Hurwicz [1986]. These papers also contain an almost exhaustive list of references. The ideas expressed here are partly based on Hahn [1982] and van der Laan [1988].

Roughly speaking, an adjustment process is a process operating in an economy to bring about an equilibrium. We can categorize the processes defined in the literature according to different criteria. In economic theory, the most important categorization distinguishes three types of adjustment processes, namely price adjustment processes, resource allocation mechanisms, and decentralistic processes based on pairwise bargaining. Let us discuss them in somewhat more detail.

The price adjustment processes are based on the idea that an equilibrium is reached by adaptations of prices. The basic idea for these processes is the well-known 'law of demand and supply' for a single good. That law states that on a market the price of a good increases if the demand for that good is larger than its supply. Similarly, its price decreases in the reverse case. By adaptations of this kind an equilibrium on a single market is reached. Walras [1874] generalized this concept to a situation with more goods. First, the law of demand and supply is applied to the first good, next to the second good, and so on. From this it is only a small step to define a similar process by which the same adaptations simultaneously take place on a number of individual markets. This formulation is known as the Walrasian adjustment process and has been introduced by Samuelson [1947]. Later on several generalizations of the latter process have been introduced. The most important one is the Newton-like method of Smale [1976], in which there are also dependencies among the markets, i.e. the price adaptation on one market is also determined by the situation on the other markets. The economic interpretation of Smale's method is that the price adaptations are such that both the excess demands and the excess supplies are simultaneously and proportionally diminished.

This might not be the case for the Walrasian process. Consider for example the market for a Giffen good. If such a market reveals an excess demand then the resulting price increase by the Walrasian process will increase this excess.

The price adjustment processes mentioned thusfar are tâtonnement processes. The crucial feature of such processes is that trades are only based on final values. Prices are adjusted till their equilibrium values are attained. Then trade takes place at these equilibrium prices. On the other hand we have the nontâtonnement price adjustment processes. In those processes trade also takes place during the adjustment procedure. The difficulty is of course to model the trade at these non market clearing prices. For this trade the assumption of orderly markets is usually made. A market is orderly if the short side of the market is always satisfied. For example, when a consumer reveals a demand, that demand is satisfied if the market has an excess supply.

The price adjustment processes as defined above are centralistic in the sense that they all assume the existence of an auctioneer to adapt the prices. Consider for example the Walrasian process. Its performance can be thought of as follows. At a certain price vector all consumers reveal their wishes to the auctioneer. The auctioneer collects the wishes and after comparing the total demand for and the total supply of all goods he prescribes a new vector of prices, and so on. Similarly, Smale's process can be seen in this way. Also the nontâtonnement processes need an auctioneer. Before at a price vector trade can take place an auctioneer must consider the totals of demand and supply on each market.

Finally, we remark that in general price adjustment processes can be formulated both as continuous procedures and as discrete procedures. Note that these two possible ways relate to different treatments of time. At first sight this distinction may seem to be academic, but it appears that things like convergence may heavily depend on this. We return to this point later on.

In the sixties, Hurwicz [1960] and others designed the concept of resource allocation mechanisms. This is a much broader concept and includes price adjustment processes. Crucial notions here are messages and response functions. Every agent in the economy sends a message. Based on these messages agents send in new messages till nobody wants to change. At these equilibrium messages trade takes place. This is a general description of a tâtonnement mechanism. But here we can also distinguish non-tâtonnement mechanisms. An important class of such mechanisms are the strategic market games as studied for example by Shubik [1984, ch. 15]. As already indicated the mechanism concept is very broad. It not only encompasses competition, but it is also suited for other market and nonmarket conditions. Examples of exchange processes without prices that fit in the mechanism concept are the Edgeworth process (see Uzawa [1962]) and the bidding process of Hurwicz, Radner, and Reiter [1975]. Again, most of the mechanisms given here are centralistic in the sense as discussed before. Important questions in this research area are for example 'Which mechanism utilizes the least information to establish certain features', or 'Design a mechanism to perform a specified social goal'. A drawback of resource allocation mechanisms is that their message space is usually much larger than the prices needed for the price adjustment processes. The costs for implementation are therefore also larger.

Finally, we briefly discuss the decentralistic processes based on pairwise bargaining. This type of processes has been initiated by the article of Rubenstein [1982]. Here the trade and price formation occurs via pairwise bargaining. More concrete, assume an economy with a large number of agents. Two agents meet and bargain about a transaction. If there is agreement they trade, else they search for another trader. Eventually the economy reaches a stationary state, the equilibrium. Under certain conditions this equilibrium state appears to be the Walrasian equilibrium (see for example Gale [1986]).

The usefulness of the concept of adjustment processes and the contributions given above is strongly related to the importance of equilibrium analysis. The latter has appeared to be fruitful for example to study policy implications by comparative statics. There, the result of a given policy is measured as the difference between two equilibrium states. The existence of adjustment processes revealing natural intrinsic forces in the economy which drive that economy towards an equilibrium state gives an extra argument for that approach. In this context we also should mention the occurence of multiple equilibria (see Kehoe [1985]). It is plausible that the existence of an adjustment process may decide upon the question which equilibrium is most likely to occur. This because the convergence of the process to that specific equilibrium indicates that the equilibrium is stable.

Another question concerns the problem whether the processes defined thusfar correspond to what happens in reality. In fact, the ultimate goal of the study of adjustment processes is to elucidate the mechanisms and forces that tend to bring an economy towards an equilibrium. We think that in this respect the state of the art is just in the beginning. Most of the processes in the literature are centralistic and need an auctioneer. This is not very convincing. The same holds for the tâtonnement processes. In practice we do see exchanges against non-clearing prices. We think that an adjustment process which resembles reality should be decentralistic. In that sense the research along the lines of Rubenstein seems to be the most promising. On the other hand, Rubenstein requires rational agents willing to bargain with everyone for every commodity. Economic agents, however, are making use of institutions, such as markets, and for good reasons. Thus, the practical worth of these processes may be limited.

Concluding we may say that the theory concerning adjustment processes developed thusfar has not so much value as a description of reality. But the theory is very important for general equilibrium analysis, both empirically and theoretically. For such analysis the convergence of an adjustment process is of crucial importance. This monograph pretends to yield a contribution to this field. We consider a new type of tâtonnement price adjustment processes in which the adjustments are governed by the excess demands and the location of the initial price vector. The advantage of our processes above the other ones developed until now is that they converge under far weaker conditions. Besides, they are rather easy to adapt for application to a broader range of models. For example, in this monograph we consider applications to an international trade model and to an exchange economy with linear production.

Let us consider now the convergence issue in more detail. Here we follow the lines of Saari and Simon [1978] and Saari [1985]. They consider the convergence of tâtonnement price adjustment processes in the context of a pure exchange economy. From Debreu [1974] we know that every continuous function $z: \mathbb{R}^{n+1}_{++} \to \mathbb{R}^{n+1}$ from the price space into the commodity space, satisfying $\Sigma_i p_i z_i(p) = 0$, can be thought of as a representation of a specific exchange economy with n+1 goods, indexed from 1 to n+1. Here $p = (p_1,\ldots,p_{n+1})^{\top}$ in $\mathbb{R}^{n+1}_{++}$ represents a strictly positive price vector whereas z(p) is the (n+1)-vector of excess demands at price vector p. The i-th component of z(p), denoted $z_i(p)$, is positive (negative) whenever the market for good i is in excess demand (supply) at p. Market i is said to be in equilibrium at p if $z_i(p) = 0$. At an equilibrium price vector p* all markets are in equilibrium, i.e. $z(p^*) = \underline{0}$. Now, Saari and Simon call an adjustment process effective if there exists an open set of prices, D, in $\mathbb{R}^{n+1}_{+}$ such that for almost all economies this process converges from almost all price vectors in D towards an equilibrium price vector. Furthermore, a price mechanism is called locally effective if for almost all economies, all equilibria in such an economy have an open neighbourhood such that whenever the process starts from a price vector in this neighbourhood it converges towards that equilibrium.

Saari and Simon [1978] considered the existence of (locally) effective adjustment processes defined as a differential equation, and Saari [1985] did the same for iterative procedures. Their conclusions were that any continuous price mechanism based on only information about z(p) and its derivatives at every price vector p, requires at least information concerning z(p) and the jacobian matrix Dz(p) in order to be effective or locally effective. The same holds for any locally effective iterative procedure. This does not say that there indeed exist locally effective procedures. The required amount of information is a necessary condition, not a sufficient one. It can be shown that there exists no effective iterative price adjustment procedure based on a finite amount of information. These results are very disappointing. First of all, because (locally) effective mechanisms need a lot of information. Secondly, the nonexistence of an effective iterative procedure based on only information about z(p) and its derivatives is very troublesome because also the price paths generated by the continuous procedures have in practice to be followed by a discrete procedure. The processes we consider in this monograph are continuous procedures in which the adaptations of a given price vector are governed not only by the related vector of excess demands, but also by the price vector from which the whole procedure started. The latter feature is crucial and guarantees that the processes reach an equilibrium from almost any price vector in the price space. Thus, our processes are effective in a strong sense. We call these processes globally effective.

We conclude this section by considering some of the price adjustment processes for a pure exchange economy in somewhat more detail against the background given above. First of all we have the law of demand and supply applied to a single market. Formulated as a differential equation its convergence is guaranteed. In general this is not the case for its iterative counterpart. When successively applied to more markets the law of supply and demand might not be converging since an equilibrium on one market can be disturbed by adaptations of a price on another market. The most well-known adjustment process is the Walrasian price adjustment process. It follows the curve of solutions to the differential equation $\dot{p} = z(p)$. However, strong assumptions on the excess demand function are needed to guarantee convergence. For the Walrasian procedure to be effective all goods need to be gross substitutes or a strong revealed preference assumption must hold. To guarantee local effectiveness these assumptions can be weakened somewhat to for example diagonal dominance of the jacobian matrix at the equilibrium, i.e. at an equilibrium the absolute value of the own price effect exceeds the sum of the absolute cross effects. Scarf [1960] gives some examples of excess demand functions for which the Walrasian tâtonnement process fails to converge from any starting point other than the equilibrium. The reason why this is the case for Scarf's famous three-goods-example has been clarified by Keenan and Rader [1985]. They discovered a necessary and sufficient condition on an excess demand function for a three-goods-economy, such that the Walrasian process is globally effective. This condition, stating that the sum of the diagonal elements of $Dz(p)$ must be negative for all p, is not fulfilled in the Scarf-example.

The global Newton method of Smale [1976] (see also Keenan [1981]) follows the solution curve to the differential equation $Dz(p)\dot{p} = -\lambda(p)z(p)$, where $\lambda(p)$ is related to the sign of the determinant of $Dz(p)$. This procedure is both effective and locally effective, and as indicated by Saari and Simon [1978] it indeed only needs information on p and $Dz(p)$.

The set D corresponding to the effectiveness is an open set of the boundary of the price space, where additionally an extra condition concerning Dz has to be fulfilled.

The processes to be introduced in this monograph adapt the prices according to the sign pattern of the excess demand vector and the location of the starting price vector. In Section 1.2 we give a somewhat more elaborate treatment. Van der Laan and Talman [1987a] state that these processes can be written as sequences of differential equations. As already indicated the processes are globally efficient. This is due to the fact that during their operation the starting price vector is kept in mind. This makes that the processes utilize some global information. This is different from the other procedures which only work with local information related to the ongoing price vector.

1.2. Some examples of globally efficient processes

In this section we give a first idea of the adjustment processes and algorithms we will discuss in this monograph. In the first part we present a price adjustment process for an exchange economy which has been presented by van der Laan and Talman [1987a]. It serves as an introduction to the processes to be presented in this monograph. Next, we discuss a process for finding a Nash equilibrium in a bi-matrix game and show that it can be interpreted as a strategy adjustment process. In this chapter we want to circumvent the use of many formulas. The set-up will therefore be rather intuitive.

To introduce the price adjustment process we first consider Figure 1.2.1. In this figure the excess demand pattern for an exchange economy with three goods is given. The prices have been normalized to sum up to one. The three curves indicate the price vectors at which one of the goods is in equilibrium (its excess demand is zero). For each i, $z_i(p) = 0$ divides the price set in two parts. In the part which includes the vector with $p_i = 1$, we have that $z_i(p) < 0$. For the part which includes the prices p with $p_i = 0$ we have that $z_i(p) > 0$. The curves intersect at p^*, the equilibrium price vector in this economy. Now, let us consider a price adjustment process starting from a price vector v that lies in the neighbourhood of the vector (0,1,0), i.e. at v the prices of commodities 1 and

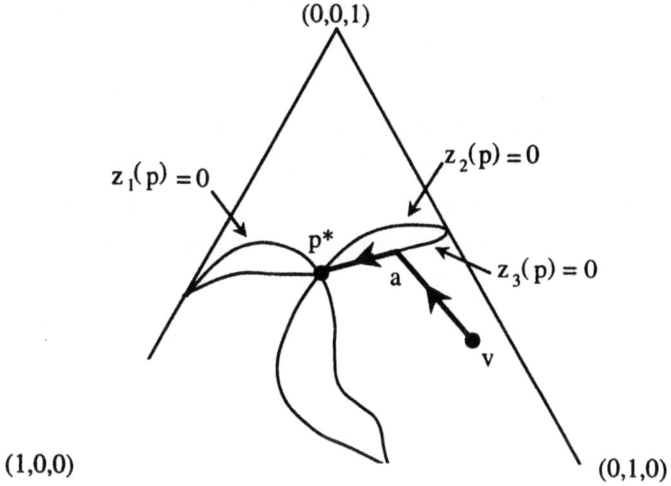

Figure 1.2.1. The price space of an exchange economy with three goods consists of price vectors $p = (p_1,p_2,p_3)$ in $\mathbb{R}^3_+$ such that $p_1+p_2+p_3 = 1$.

3 are relatively low, whereas the price of commodity 2 is relatively high (see Figure 1.2.1). For this price vector, it seems reasonable that the commodities 1 and 3 are in excess demand, whereas good 2 is in excess supply.

The process now increases the prices of goods 1 and 3 proportionally equally while the price of commodity 2 is decreased, keeping the sum of the prices equal to one. In the figure this means that the process leaves v into the direction opposite to (0,1,0). The process continues in this way till it reaches the price vector a on the curve of prices p with $z_3(p) = 0$. At that price vector commodity 3 becomes in equilibrium, while commodity 1 is still in excess demand and commodity 2 in excess supply. From the price vector a onwards, commodity 3 is kept in equilibrium by allowing its relative price to become below the relative price of good 1. Here the relative price of a good is its price relative to its starting price, i.e. with the relative price of i at p we mean p_i/v_i. Recall that these relative prices of the goods 1 and 3 are equal along the line segment between v and a. So, the process moves from a along the curve at which good 3 is in equilibrium, towards p*, because in that direction the relative price of good 3 becomes smaller than the relative price of good 1.

In general the process starts from a price vector v by increasing the prices of the goods in excess demand relatively equally while the prices of the commodities in excess supply are decreased relatively equally. The process follows a path of price vectors at which the relative prices of goods in excess demand (supply) are maximal (minimal), whereas the relative prices of goods in equilibrium may vary between the maximal and minimal value in order to keep them in equilibrium.

Next, we want to give an idea of the process for finding a Nash equilibrium in a bi-matrix game. The strategy space of a bi-matrix game in which each of the two players has two actions is depicted in Figure 1.2.2.

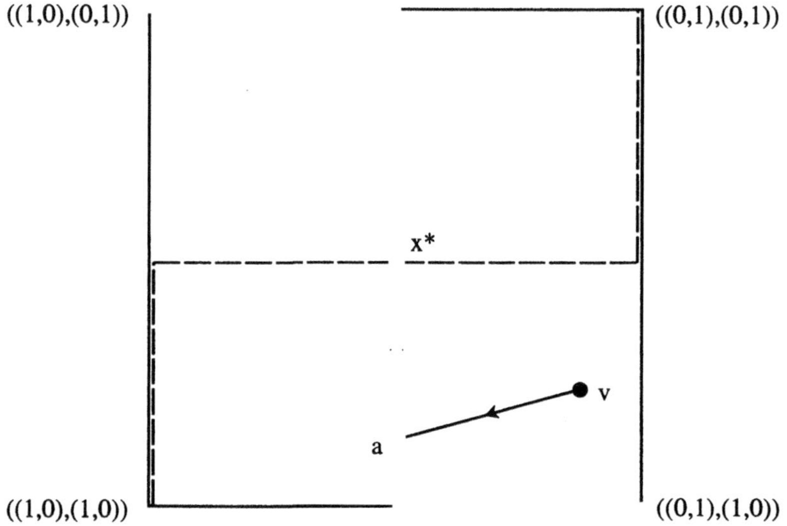

Figure 1.2.2. The strategy space of a bi-matrix game with each player having two actions. The space consists of the points $x = ((x_{11},x_{12}),(x_{21},x_{22}))$, where x_{jk} is the probability with which player j plays his k-th action, $j,k \in \{1,2\}$. For $j \in \{1,2\}$, $x_{j1} \geq 0$, $x_{j2} \geq 0$, and $x_{j1}+x_{j2} = 1$.

The thick piecewise linear curve in the figure represents the best reply set of player 2. Player 2 is indifferent between his actions if player 1 plays his actions each with probability 1/2. Furthermore, action 2 is optimal for player 2 if player 1 plays his first action with a larger probability than his second action. In the reverse case action 1 is optimal for player 2. Similarly, we can introduce the best reply set of player 1 being the dashed piecewise linear curve. At the Nash equilibrium x^* the two curves intersect and both players play optimal against the mixed strategy of their opponent.

Let us now consider the process starting from the strategy vector v. The path followed by the process is indicated in the figure and consists of two parts; the segment between v and a, and the segment from a to x^*. In practice this is done by complementary pivoting and x^* is reached in two steps. However, the path can be interpreted as being generated by a strategy adjustment procedure. Consider the strategy vector v. From the best reply curves we deduce that at v it is optimal for both players to play their first action. Then the algorithm increases the probabilities with which these actions are played whereas the probabilities related to the second, nonoptimal, action of both players are decreased proportionally. Thus, in the figure the procedure moves towards ((1,0),(1,0)) where both players play their first action with probability 1. But at the vector a, player 2 becomes indifferent between his actions. Both his actions are optimal and we say that the player 2 is in equilibrium. Now, from the strategy vector a onwards player 2 is kept in equilibrium, i.e. the process stays on his best reply set and allows the probability related to his second action to become larger than that one related to action 2 of player 1. This because the latter action is now the only nonoptimal action. In this way the process reaches the Nash equilibrium x^*. So, the main idea of the algorithm is to drive down to zero the probabilities with which nonoptimal actions are played. If they are zero a Nash equilibrium has been reached.

1.3. Outline of the monograph and main results

In Chapter 2 we give the mathematical background needed in the remainder of this monograph. Besides, we introduce some notational conventions. All the other chapters can be studied independently from each other. The manuscript in fact consists of two parts. The first part (Chapters 3, 4 and 5) deals with price adjustment processes. The second part (Chapter 6) deals with algorithms for solving noncooperative games.

Chapter 3 is rather fundamental and studies conditions for the existence and convergence of our price adjustment processes. The techniques used are the homotopy method and transversality theory. Besides, this chapter contains a new and very appealing price adjustment process. Along the path of that process the excess demands are a nonnegative multiple of the difference between the related price and starting price.

In the Chapters 4 and 5 we generalize the price adjustment process for a pure exchange economy for application to other models. Chapter 4 deals with a process for an international trade model, whereas in Chapter 5 we present an adjustment process for an exchange economy with linear production. Thusfar only algorithms for finding an equilibrium in such economies have been developed, but never an adjustment procedure. The adjustment procedure in Chapter 4 adapts prices and exchange rates whereas the process in Chapter 5 adapts prices and activity levels.

In Chapter 6 we focus on game theory. In the first half of that chapter we consider an algorithm for finding a Nash equilibrium in a bi-matrix game. Next, we give a game-theoretic interpretation of the Lemke-Howson method, which is the standard method for finding such an equilibrium, and compare the two methods. Our algorithm is superior in the sense that it can find Nash equilibria which cannot be found by the Lemke-Howson method. Interesting is its game-theoretic interpretation and the fact that it always finds a perfect equilibrium whenever it starts from a completely mixed strategy vector. Finally, we give a generalization of our algorithm for application to noncooperative games with more than two players. We approach a Nash equilibrium for such a game by a repeated application of the algorithm for the bi-matrix game generalized for the linearized more-person game. For each linearized game we find an exact solution which can be considered as an approximate Nash equilibrium for the original game.

CHAPTER 2

Preliminaries

In this chapter we review some mathematical concepts which are needed in the sequel. The chapter is divided into four parts. In Section 2.1 we introduce some notational conventions and present the definitions of a unit simplex and a simplotope. The latter two are the sets on which almost all problems considered later on are defined. Section 2.2 deals with the stationary point problem. In fact, all the problems considered in this monograph can be viewed upon in this manner. In Section 2.3 we consider some notions being related to simplicial algorithms. Such algorithms serve to find a stationary point of an arbitrary continuous function. In later chapters we will see how these algorithms are related to the processes considered in this monograph. Finally, in Section 2.4 we discuss some concepts from differential topology. These notions are of importance when we consider the existence of the adjustment processes.

2.1. Notation

The set of real numbers is denoted by $\mathbb{R}$, whereas $\mathbb{R}_+$ indicates the set of all nonnegative real numbers. Let k be a given positive integer. Then the set $\mathbb{R}^k$ denotes the k-dimensional Euclidean space. An element $x \in \mathbb{R}^k$, also called a vector or point, is a k-tuple of real numbers and can be written as $x = (x_1, x_2, \ldots, x_k)^\top$, with $x_i \in \mathbb{R}$, $i \in \{1, \ldots, k\}$, called the i-th component of x. Here the symbol $\top$ is used to indicate the transpose. Given two vectors x,y in $\mathbb{R}^k$, we write $x > y$ if x is larger than y in all components. By $x \geq y$ we mean that at least one component of x is larger than the corresponding component of y, whereas the other components of x are at least equal to those of y. If x is at least as large as y in all components we write $x \geqq y$. Accordingly we define $<$, $\leq$, and $\leqq$. Related to two vectors x,y in $\mathbb{R}^k$, $[x,y]$ denotes the line segment connecting x and y including x and y. When the vectors x and y are excluded we denote (x,y).

Similarly, we denote $[x,y)$ and $(x,y]$ when y resp. x is excluded. We frequently indicate certain subsets of $\mathbb{R}^k$ by a specific symbol. The nonnegative orthant of $\mathbb{R}^k$ is denoted by $\mathbb{R}^k_+$. Thus, $\mathbb{R}^k_+ := \{x \in \mathbb{R}^k | x_i \geqq 0,\ i \in \{1,\dots,k\}\}$. Similarly, we have $\mathbb{R}^k_- := \{x \in \mathbb{R}^k | x_i \leqq 0,\ i \in \{1,\dots,k\}\}$ and $\mathbb{R}^k_{++} := \{x \in \mathbb{R}^k | x_i > 0,\ i \in \{1,\dots,k\}\}$. The set $\mathbb{R}^k$ is endowed with the standard metric derived from the Euclidean norm $\|\cdot\|$, defined by $\|x\| = (\Sigma^k_{i=1} x_i^2)^{\frac{1}{2}}$, $x \in \mathbb{R}^k$. The inner product of two vectors $x,y \in \mathbb{R}^k$ is denoted by $x \cdot y$. The i-th unit vector of $\mathbb{R}^k$ is denoted by $e^k(i)$, $i \in \{1,\dots,k\}$, whereas the vector of ones in $\mathbb{R}^k$ is denoted by e^k. The vector of zeros in $\mathbb{R}^k$ is denoted by $\underline{0}^k$. When the dimension of a vector is clear from the context we often delete the superscript. The expression 'k-dimensional set (or vector)' is often abbreviated to k-set (k-vector). The unit matrix is denoted by E and by $|A|$ we denote the number of elements in a finite set A, whereas $I_k := \{1,\dots,k\}$. Finally, given a set $A \subset \mathbb{R}$, max A (min A) denotes the maximum (minimum) element of A (if it exists).

A notion frequently used in the sequel is that of a convex hull of a set.

<u>Definition 2.1.1</u>. The *convex hull* of a set C in $\mathbb{R}^k$, denoted conv(C), consists of all the points which are convex combinations of points in C, i.e.

$$\text{conv}(C) = \{x \in \mathbb{R}^k | x = \Sigma^h_{i=1} \lambda_i x^i, \text{ with } \Sigma^h_{i=1} \lambda_i = 1, \text{ and for all}$$

$$i \in I_h,\ \lambda_i \in \mathbb{R}_+ \text{ and } x^i \in C\}.$$

The affine hull of a set C in $\mathbb{R}^k$, aff(C), is defined in the same way except that the condition $\lambda_i \in \mathbb{R}_+$ is weakened into $\lambda_i \in \mathbb{R}$ for all i. We note that in this monograph the boundary, closure, and interior of a set C - notation: bd(C), cl(C), and int(C)- are always defined relative to aff(C).

Two kinds of sets in the Euclidean space are of extreme importance in this monograph. These are the unit simplex and the simplotope. Let n be a given positive integer. The n-dimensional unit simplex is the set $\{x \in \mathbb{R}^{n+1}_+ | \Sigma^{n+1}_{i=1} x_i = 1\}$ and is denoted by S^n. Clearly, S^n is the convex hull of the n+1 unit vectors in $\mathbb{R}^{n+1}$. Further, $\text{aff}(S^n)$ is the set $\{x \in \mathbb{R}^{n+1}|$

$\Sigma_{i=1}^{n+1} x_i = 1\}$. In Figure 2.3.1, $\mathrm{aff}(S^2)$ is the plane. Henceforth, $\mathrm{int}(S^n)$ is equal to $\{x \in S^n | x_i > 0,\ i \in I_{n+1}\}$, whereas $\mathrm{bd}(S^n)$ equals $\{x \in S^n | x_i = 0$ for some $i \in I_{n+1}\}$. A simplotope is the Cartesian product of several unit simplices and is denoted by S. More precisely, for given positive integers $n_1,\ldots,n_n$, $S := \Pi_{j=1}^n S^{n_j} = \{x = (x_1,\ldots,x_n) | x_i \in S^{n_i},\ i \in \{1,\ldots,n\}\}$. The set S^n, $n = 2$, is illustrated in Figure 2.3.1, whereas in Figure 2.3.2 the set $S = S^1 \times S^1$ is drawn. Observe that both S^2 and $S^1 \times S^1$ are 2-dimensional sets although S^2 is a subset of $\mathbb{R}^3$ while $S^1 \times S^1$ lies in $\mathbb{R}^4$. We remark that both S^n and S are convex and compact.

Related to S^n and S we introduce some notation. $S^n(T)$, $T \subset I_{n+1}$, denotes the set $\{x \in S^n | x_i = 0,\ i \notin T\}$. Thus, in particular, $S^n(\{i\}) = \{e(i)\}$, $i \in I_{n+1}$. To denote $S^n(I_{n+1} \setminus \{i\})$, $i \in I_{n+1}$, we often use the notation S_i^n. Concerning S we call an element $x = (x_1,\ldots,x_n)$ in S also a vector or point in S. The number x_{jk} denotes the k-th component of x_j and is also called the (j,k)-th component of x, $k \in \{1,\ldots,n_j+1\}$. The index set $\{(j,1),(j,2),\ldots,(j,n_j+1)\}$ related to S^{n_j} is denoted by $I(j)$, whereas $I := \cup_{j=1}^n I(j)$. By $S(T)$, $T \subset I$ with $T(j) := T \cap I(j) \neq \emptyset$ for all $j \in I_n$, we denote the set $\{x \in S | x_{ih} = 0,\ (i,h) \notin T\}$. In particular, $S(T^0)$, with $|T^0 \cap I(j)| = 1$ for all $j \in I_n$, is equal to $\{e(T^0)\}$, with $e(T^0)$ being the vector in S for which $e_{ih}(T^0) = 1$, $(i,h) \in T^0$, and $e_{ih}(T^0) = 0$ elsewhere. Observe that the simplotope S is the convex hull of all these vectors $e(T^0)$. Furthermore, $\mathrm{int}(S)$ equals $\{x \in S | x_{ih} > 0,\ (i,h) \in I\}$ and $\mathrm{bd}(S) = \{x \in S | x_{ih} = 0$ for some $(i,h) \in I\}$. In general, $e(T)$, $T \subset I$, denotes the vector in $\Pi_{j=1}^n \mathbb{R}^{n_j+1}$ for which $e_{ih}(T) = 1$, $(i,h) \in T$, and $e_{ih}(T) = 0$ elsewhere. Finally, for $(i,h) \in I$ we denote the set $S(I \setminus \{(i,h)\})$ by S_{ih}.

We conclude this section by introducing the notion of a sign vector and some related notation. A sign vector s in $\mathbb{R}^k$ is a vector whose components s_i, $i \in I_k$, lie in $\{-1,0,+1\}$. Related to a sign vector s in $\mathbb{R}^k$ we define the following subsets of I_k:

$$I^0(s) = \{j \in I_k | s_j = 0\}$$

$$I^+(s) = \{j \in I_k | s_j = +1\}$$

$$I^-(s) = \{j \in I_k | s_j = -1\}.$$

Similarly, a sign vector s in $\prod_{j=1}^n \mathbb{R}^{n_j+1}$ is a vector whose components s_{ih}, $(i,h) \in I$, lie in $\{-1,0,+1\}$. Related to such a vector s, for each $h \in I_n$ we define

$$I_h^0(s) = \{(h,k) \in I(h) | s_{hk} = 0\}$$

$$I_h^+(s) = \{(h,k) \in I(h) | s_{hk} = +1\}$$

$$I_h^-(s) = \{(h,k) \in I(h) | s_{hk} = -1\}.$$

By $I^0(s)$, $I^+(s)$, and $I^-(s)$ we denote the union of the corresponding subsets over all h. The sign vector related to a vector y in $\mathbb{R}^k$, notation sgn(y), is a vector s in $\mathbb{R}^k$ with components in $\{-1,0,+1\}$, where for $i \in I_k$, $s_i = +1$ (-1) if $y_i > 0$ (< 0), while $s_i = 0$ if $y_i = 0$. Similarly, sgn(y), $y \in \prod_{j=1}^n \mathbb{R}^{n_j+1}$, is a vector s in $\prod_{j=1}^n \mathbb{R}^{n_j+1}$ with for all $(i,h) \in I$, $s_{ih} = +1$ (-1) if $y_{ih} > 0$ (< 0), while $s_{ih} = 0$ if $y_{ih} = 0$.

2.2. The stationary point problem

This section is mainly devoted to the theorem stating that every continuous function defined on a nonempty, convex, and compact set has a stationary point. It turns out that this very general existence theorem underlies the existence proofs for equilibria in games and exchange economies. Furthermore, we treat some properties of the function involved. In fact we give here a mathematical framework which returns in all applications. The precise meaning and interpretation of the properties become clear in the context of the application involved.

Definition 2.2.1. Let C be a subset of $\mathbb{R}^k$ and let $f : D \to \mathbb{R}^k$ be a function, with D a subset of $\mathbb{R}^k$ containing C. A point x^* in C is a *stationary point* of f on C if

$$x \cdot f(x^*) \leqq x^* \cdot f(x^*) \text{ for all } x \text{ in } C. \qquad (2.2.1)$$

Lemma 2.2.2. Let $f : D \to \mathbb{R}^k$, $D \subseteq \mathbb{R}^k$, be a continuous function and let C be a nonempty, convex, and compact subset of D. Then f has at least one stationary point on C.

The proof of the theorem can be found in Doup [1988, p. 24] and goes via a fixed-point argument. In fact, the theorem is equivalent with Brouwer's fixed point theorem stating that every continuous function $f : C \to C$, with C in $\mathbb{R}^k$ nonempty, compact, and convex, has a fixed point x^*, i.e. $f(x^*) = x^*$ (see Brouwer [1912]). In the sequel, for the set C we mostly take the unit simplex S^n or a simplotope S and we also impose some specific conditions on f. Under these conditions on f and C we can derive additional properties to hold at a stationary point. Let us first treat the case in which C equals $S = \Pi_{j=1}^n S^{n_j}$ and $f : S \to \Pi_{j=1}^n \mathbb{R}^{n_j+1}$. In that case for $x \in S$, $x \cdot f(x)$ equals $\Sigma_{j=1}^n x_j \cdot f_j(x)$.

Definition 2.2.3. Let f be a function from S to $\Pi_{j=1}^n \mathbb{R}^{n_j+1}$. The function f satisfies the *complementarity condition* if for all $x \in S$

$$x_j \cdot f_j(x) = 0 \text{ for all } j \in I_n. \tag{2.2.2}$$

The function f satisfies the *boundary condition* if for all $x \in S$

$$f_{ih}(x) \geqq 0 \text{ when } x_{ih} = 0,\ (i,h) \in I. \tag{2.2.3}$$

Theorem 2.2.4. Let $f : S \to \Pi_{j=1}^n \mathbb{R}^{n_j+1}$ be continuous and let x^* be a stationary point of f on S. Then the following statements hold:

A.
$$\begin{aligned} f_{jk}(x^*) &= \max_\ell f_{j\ell}(x^*) && \text{if } x^*_{jk} > 0 \\ f_{jk}(x^*) &\leqq \max_\ell f_{j\ell}(x^*) && \text{if } x^*_{jk} = 0. \end{aligned} \tag{2.2.4}$$

B. If f satisfies the complementarity condition (2.2.2) then

$$f_{jk}(x^*) = 0 \qquad \text{if } x^*_{jk} > 0$$
$$f_{jk}(x^*) \leqq 0 \qquad \text{if } x^*_{jk} = 0. \tag{2.2.5}$$

C. If (2.2.2) and the boundary condition (2.2.3) hold then

$$f(x^*) = \underline{0}. \tag{2.2.6}$$

D. If (2.2.2) holds and (2.2.3) holds with strict inequality then

$$f(x^*) = \underline{0} \text{ and } x^* \in \text{int}(S). \tag{2.2.7}$$

Proof. Statement A follows for indices in I(j) from substituting for x in (2.2.1) respectively the n_j+1 vectors x^ℓ in S with $x^\ell_i = x^*_i$, $i \neq j$, and $x^\ell_j = e(\ell)$, $\ell \in \{1,\ldots,n_j+1\}$. The other statements follow directly by combining (2.2.4) and Definition 2.2.3.

□

Observe that statement A is equivalent with the definition of a stationary point. The corresponding definition and theorem for the special case that S equals S^n follow straightforward. For convenience and later reference we present the analogues of Definition 2.2.3 and Theorem 2.2.4.

Definition 2.2.5. Let f be a function from S^n to $\mathbb{R}^{n+1}$. The function f satisfies the *complementarity condition* if

$$x \cdot f(x) = 0 \text{ for all } x \in S^n. \tag{2.2.8}$$

The function f satisfies the *boundary condition* if

$$f_i(x) \geqq 0 \text{ when } x_i = 0,\ i \in I_{n+1}. \tag{2.2.9}$$

Theorem 2.2.6. Let $f : S^n \to \mathbb{R}^{n+1}$ be continuous and let x^* be a stationary point of f on S^n. Then the following statements hold:

A. $f_i(x^*) = \max_k f_k(x^*)$ if $x_i^* > 0$

$f_i(x^*) \leqq \max_k f_k(x^*)$ if $x_i^* = 0$. (2.2.10)

B. If f satisfies the complementarity condition (2.2.8) then

$f_i(x^*) = 0$ if $x_i^* > 0$

$f_i(x^*) \leqq 0$ if $x_i^* = 0$. (2.2.11)

C. If (2.2.8) and the boundary condition (2.2.9) hold then

$$f(x^*) = \underline{0}. \quad (2.2.12)$$

D. If (2.2.8) holds and (2.2.9) holds with strict inequality then

$$f(x^*) = \underline{0} \text{ and } x^* \in \text{int}(S^n). \quad (2.2.13)$$

We conclude this section with some terminology and show that well-known problems fit in the framework given above. The problem of finding a stationary point of a function f on a set C is called the stationary point problem (SPP) of f on C. The SPP results in the nonlinear complementarity problem (NLCP) when f satisfies the complementarity condition (2.2.2). The linear complementarity problem (LCP) of f on S is the SPP of f on S with f satisfying (2.2.2) and linearity. The zero point problem (ZPP) of f on S is the problem of finding a zero point of f on S. Under conditions (2.2.2) and (2.2.3) on f, the SPP of f on S boils down to a ZPP (see Theorem 2.2.4). From Definition 2.2.1 it is obvious that zero points are stationary points.

<u>Example 2.2.1</u>. (see Scarf [1960, p. 162]) Consider the function $f : S^2 \to \mathbb{R}^3$ given by

$$f(x) = (-x_2+x_3, -x_3+x_1, -x_1+x_2)^\top.$$

Observe that f is continuous and satisfies the complementarity condition but not the boundary condition. Because of the latter, for a stationary point x^* must hold $f(x^*) \leqq \underline{0}$. From this it is easy to derive that the only stationary point is $x^* = (1/3,1/3,1/3)^T$ with $f(x^*) = \underline{0}$.

2.3. Notions related to simplicial algorithms

The central concept in this section is that of a subdivision. At the end of this section we briefly illustrate the relation between a subdivision and a simplicial algorithm. But before giving the definition of a subdivision we need to define some other concepts.

Definition 2.3.1.. A *polytope* σ in $\mathbb{R}^k$ is the convex hull of a finite number of points in $\mathbb{R}^k$. A subset τ of σ is called a *face* of σ if τ is a polytope and for any $p \in \tau$ and any pair $\{x,y\} \subset \sigma$ such that $p = \lambda x + (1-\lambda)y$ with $\lambda \in (0,1)$, holds that $\{x,y\} \subset \tau$. A *facet* $\bar{\tau}$ of σ is a face such that $\dim(\bar{\tau}) = \dim(\sigma) - 1$. A face of dimension 0 is a *vertex* of σ.

Observe that S^n and S are both polytopes. Concerning S^n we have that $S^n(T)$ is a (t-1)-face, with $t = |T|$. The n+1 vertices of S^n are the unit vectors in $\mathbb{R}^{n+1}$ whereas the set S^n_i is the facet of S^n opposite vertex $e(i)$, $i \in I_{n+1}$. In the sequel we often use the term cell instead of polytope, although the first concept is more general. In fact, polytopes are bounded cells. Furthermore, note that proper faces of polytopes are also polytopes but of lower dimension.

Definition 2.3.2. A *subdivision* of a k-dimensional convex and compact subset C of $\mathbb{R}^n$ is a collection G of k-dimensional polytopes, k-polytopes, such that

i) C is the union of all polytopes in G

ii) the intersection of two polytopes is either empty or a common face

iii) each facet of a k-polytope $\sigma \in G$ either lies in bd(C) and is only a facet of σ, or it does not lie in bd(C) and is also a facet of exactly one other polytope $\sigma' \in G$.

Examples of subdivisions of S^n and S are given in the Figures 2.3.1 and 2.3.2, respectively.

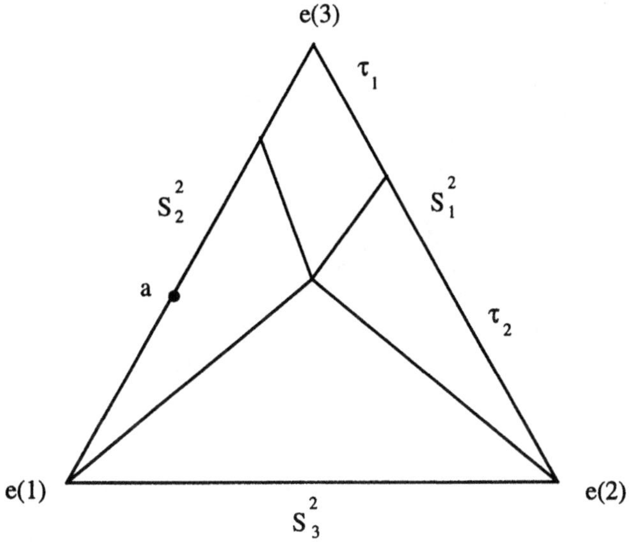

Figure 2.3.1. Subdivision of S^2.

For later reference we state the next property of a subdivision.

Lemma 2.3.3. Let G be a subdivision of a k-dimensional convex, compact subset C of $\mathbb{R}^n$. Furthermore, let D be a (k-1)-dimensional subset in bd(C) such that $D = C \cap \text{aff}(D)$. Then D is subdivided into (k-1)-polytopes which are contained in D and are facets of polytopes in G.

To illustrate the lemma, again consider Figure 2.3.1. When we take D equal to S_1^2 we see that D is subdivided into τ_1 and τ_2. Observe that the boundary segment [a,e(3)] is not subdivided. Indeed, this set does not satisfy the conditions given in the lemma.

Specific examples of polytopes and subdivisions are simplices and triangulations, respectively. Because we frequently refer to the latter two concepts we give a formal definition. But for that we need to give the notion of affine independentness. We say that the points $x^1,\dots,x^h$ in $\mathbb{R}^k$

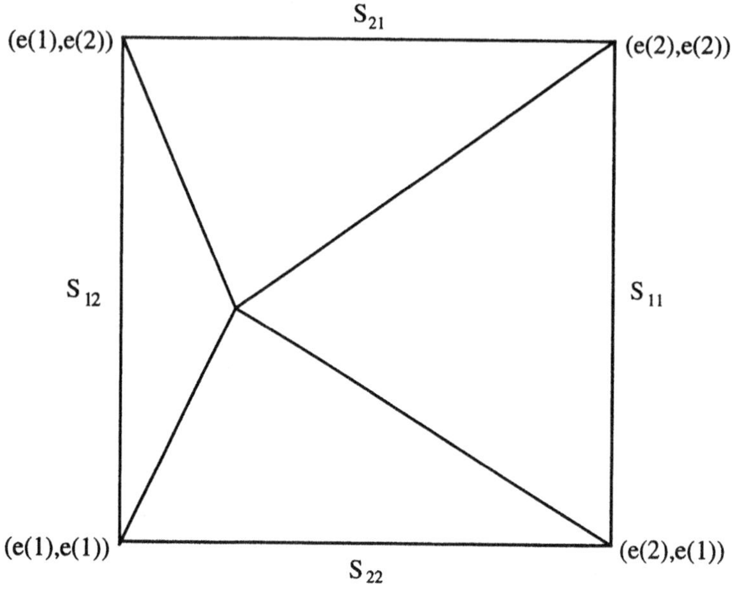

Figure 2.3.2. Simplicial subdivision of $S = S^1 \times S^1$.

are affinely independent if

$$\Sigma_{i=1}^{h} \lambda_i x^i = 0 \text{ and } \Sigma_{i=1}^{h} \lambda_i = 0 \text{ imply } \lambda_i = 0,\ i \in \{1,\ldots,h\}.$$

Note that this implies that $h \leqq k+1$.

Definition 2.3.4. A *simplex* in $\mathbb{R}^k$ is the convex hull of a finite number of affinely independent points in $\mathbb{R}^k$. A *triangulation* or *simplicial subdivision* of a convex, compact set is a subdivision of this set into simplices.

A t-dimensional simplex or t-simplex σ in $\mathbb{R}^k$, being the convex hull of the points $x^1,\ldots,x^{t+1}$ in $\mathbb{R}^k$, is denoted by $\sigma(x^1,\ldots,x^{t+1})$. The points $x^1,\ldots,x^{t+1}$ are the t+1 vertices of $\sigma(x^1,\ldots,x^{t+1})$. Observe that S^n is an n-simplex in $\mathbb{R}^{n+1}$ whereas S is not a simplex. The subdivision of S in Figure 2.3.2 is a triangulation, but the one of S^n in Figure 2.3.1 is not.

Of importance for later reference is also the notion of the mesh of a subdivision.

Definition 2.3.5. Let G be a subdivision of a nonempty, convex, and compact subset of $\mathbb{R}^k$. The *mesh* of G, mesh(G), is given by

$$\text{mesh}(G) = \max_{\sigma\in G}\max_{x,y\in\sigma}(\|x-y\|).$$

We conclude this section by discussing the concept of a piecewise linear approximation of a continuous function with respect to a triangulation.

Definition 2.3.6. Let G be a triangulation of a convex, compact, t-dimensional subset C of $\mathbb{R}^k$ and let $f : C \to \mathbb{R}^\ell$ be a continuous function. Furthermore, let x be an arbitrarily chosen point in C. Then the *piecewise linear approximation* $\bar{f}$ of f at x with respect to G is given by $\bar{f}(x) = \Sigma_{i=1}^{t+1}\lambda_i f(x^i)$, if $x = \Sigma_{i=1}^{t+1}\lambda_i x^i$, $\lambda_i \in \mathbb{R}_+$, $\Sigma_{i=1}^{t+1}\lambda_i = 1$, for $\sigma(x^1,\dots,x^{t+1})$ any simplex containing x.

Observe that $\bar{f}(x)$ is well defined because the positive λ_i's are uniquely determined. It is easily verified that $\bar{f} : C \to \mathbb{R}^\ell$ is continuous and linear on each simplex in G. We illustrate the concept with an example.

Example 2.3.1. Consider the function $f : S^2 \to \mathbb{R}^3$ defined by $f(x) = (2x_1+x_2+x_3)^{-1}(2x_1,x_2,x_3)^\top$. Furthermore, let S^2 be triangulated into the three simplices $\sigma^1(e(1),e(2),\tilde{e})$, $\sigma^2(e(1),e(3),\tilde{e})$, and $\sigma^3(e(2),e(3),\tilde{e})$, where $\tilde{e}$ denotes the barycentre $(1/3,1/3,1/3)^\top$. Following the definition, $\bar{f}$ is defined by $\bar{f}(x) = (\lambda_1+\lambda_3/2,\lambda_2+\lambda_3/4,\lambda_3/4)^\top$ if $x = \lambda_1 e(1)+\lambda_2 e(2)+\lambda_3\tilde{e} \in \sigma^1$. Similarly, we get $\bar{f}(x) = (\lambda_1+\lambda_3/2,\lambda_3/4,\lambda_2+\lambda_3/4)^\top$ if $x = \lambda_1 e(1)+\lambda_2 e(3)+\lambda_3\tilde{e} \in \sigma^2$, and $\bar{f}(x) = (\lambda_3/2,\lambda_1+\lambda_3/4,\lambda_2+\lambda_3/4)^\top$ if $x = \lambda_1 e(2)+\lambda_2 e(3)+\lambda_3\tilde{e} \in \sigma^3$. Note that $\bar{f}$ is indeed linear on each simplex.

When we apply a simplicial algorithm to find a stationary point of a continuous function on some set such a method searches first for a stationary point for the piecewise linear approximation of the function with respect to a certain simplicial subdivision of that set. By definition this stationary point is an approximation of a stationary point of the original function. By repeated application of the algorithm to simplicial subdivisions with decreasing mesh, i.e. smaller simplices, we can find a stationary point with any given accuracy. The latter is due to the fact that the piecewise linear function uniformly converges to the original function when the mesh tends to zero. The main feature of simplicial algorithms is that they converge under very weak conditions. For more details about these algorithms and triangulations we refer to Dang [1991], Doup [1988], van der Laan [1980], Talman [1980], and Todd [1976].

2.4. Concepts from differential topology

In this section we gather some notions stemming from the field of differential topology. Here we confine ourselves to those notions and theorems that are most important for us. For a general introduction on these matters we refer to Guillemin and Pollack [1974] and to Milnor [1965]. Because we only consider subsets of a Euclidean space this section is mainly based on van Geldrop [1981, ch. 2] who also confines himself to the Euclidean space. The topology we use will always be the metric topology induced by $\|.\|$. We try to illustrate the definitions below by giving some intuition. But first we introduce some notation. Given a map or function $f : M \to N$ and a set $A \subset N$ we denote by $f^{\leftarrow}(A)$ the set $\{x \in M | f(x) \in A\}$. If f has an inverse, at least restricted to A, we may also denote $f^{-1}(A)$. Furthermore, by $Df(x)$ we denote the matrix of first derivatives of f at $x \in M$ (if they exist).

Definition 2.4.1. Let $X \subset \mathbb{R}^k$ and $Y \subset \mathbb{R}^{\ell}$ be open sets. The map $f : X \to Y$ is *smooth* if all of the partial derivatives of f up to any order exist and are continuous.

More generally, one speaks about C^i-maps, $i \geq 0$, if all of the partial derivatives up to order i exist and are continuous. Smooth then corresponds to C^∞. In this monograph we only consider smooth maps. This is not a great restriction because for example in the Whitney topology (see van Geldrop [1981, p. 20]), the set of smooth maps lies dense in the set of C^2-maps. We call a set dense in some topological space $\mathcal{X}$ if it intersects every nonempty open subset of $\mathcal{X}$.

Note that Definition 2.4.1 only holds for maps defined on open sets. This because the standard derivative is only defined on open sets. However, we will also deal with maps that are defined on more general sets called manifolds. Hence, we need a counterpart of Definition 2.4.1 for maps defined on manifolds. The notion of a manifold is crucial throughout this monograph. Informally, a manifold locally looks like an open subset of the Euclidean space of corresponding dimension. For example, the surface of a ball in $\mathbb{R}^3$ is a 2-dimensional submanifold of $\mathbb{R}^3$.

Definition 2.4.2. A subset $M \subset \mathbb{R}^k$ is a *smooth submanifold* of $\mathbb{R}^k$ of dimension m if for every point $x \in M$ there exists an open neighbourhood U of x and a smooth function $\varphi : U \to \mathbb{R}^k$ such that $\varphi(x) = \underline{0}$, $\text{rank}(D\varphi(y)) = k$ for all $y \in U$, and $\varphi^{-1}(V) = U \cap M$, where $V := \{(r_1,\ldots,r_k)^\top \in \mathbb{R}^k | r_{m+1} = \ldots = r_k = 0\}$. Moreover, the pair (U,φ) is called a *submanifold chart* for M at x. If M and N are both submanifolds of $\mathbb{R}^k$ whereas $N \subset M$ then N is called a *submanifold* of M.

For example, each open subset of $\mathbb{R}^k$ is a k-dimensional submanifold of $\mathbb{R}^k$, whereas a 0-dimensional submanifold of $\mathbb{R}^k$ is a discrete set. In this monograph we frequently consider smooth 1-manifolds. A smooth 1-manifold is a collection of disjoint smooth paths and loops. Let us consider paths with two end points. They can be smoothly deformed into a closed interval. A loop is connected and bounded without boundary and can be smoothly deformed into a circle. The topology on a submanifold of $\mathbb{R}^k$ is the one induced by the topology of $\mathbb{R}^k$.

Lemma 2.4.3. A subset $M \subset \mathbb{R}^k$ is a submanifold of $\mathbb{R}^k$ of dimension m if and only if for every point $x \in M$ there exists an open neighbourhood W of x and a smooth map $\Psi : W \to \mathbb{R}^{k-m}$ such that $\text{rank}(D\Psi(y)) = k-m$ for all $y \in W$ and $\Psi^{\leftarrow}(\underline{0}) = M \cap W$.

Thus, the manifold M is locally defined as the solution set of the system $\Psi(y) = \underline{0}$, constituting k-m equations in k unknowns.

Example 2.4.1. The unit sphere $\bar{S}^{n-1} := \{x \in \mathbb{R}^n \,|\, \|x\| = 1\}$ is an (n-1)-dimensional submanifold of $\mathbb{R}^n$. This is easily seen with Lemma 2.4.3 by taking for an arbitrary point y on the sphere the map $\Psi : W \to \mathbb{R}^1$, W being an open neighbourhood of y, defined by $\Psi(x) = \|x\| - 1$, $x \in W$. In Figure 2.4.1 we illustrate Definition 2.4.2 for the unit sphere in $\mathbb{R}^2$.

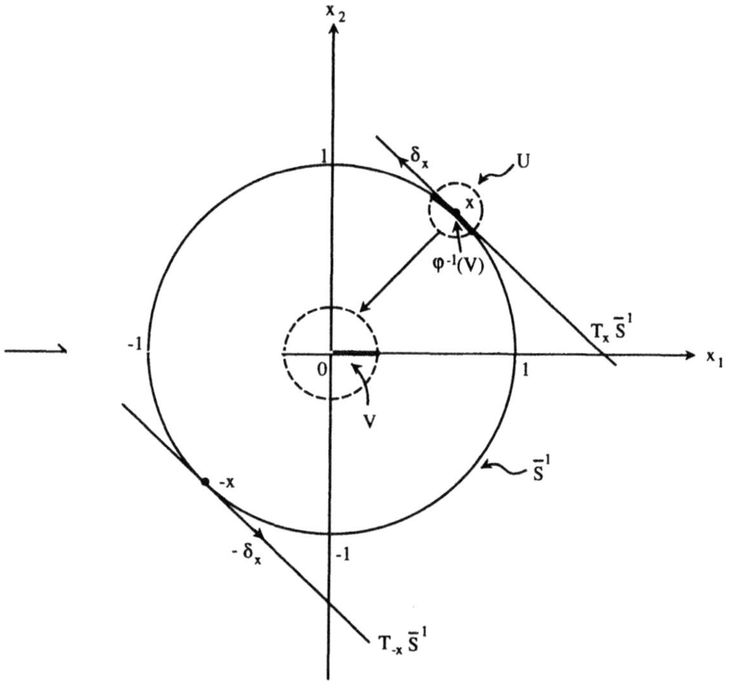

Figure 2.4.1. The set $\{x \in \mathbb{R}^2 \,|\, \|x\| = 1\}$ is a 1-submanifold in $\mathbb{R}^2$.

Next, for each x in a submanifold M we define T_xM to be the tangent space to M at x. Loosely speaking, T_xM is the linear vector space that best approximates M at x.

<u>Definition 2.4.4</u>. Let $M \subset \mathbb{R}^k$ be an m-dimensional submanifold. Further, let x be a point in M and (U,φ) a submanifold chart for M at x. Then T_xM is the set of pairs $(x,\delta y)$, where $\delta y \in D\varphi^{-1}(\underline{0})(V)$, with V as in Definition 2.4.2. The set T_xM is called the *tangent space* to M at x.

From Definition 2.4.4 we derive that T_xM has the structure of an m-dimensional vector space isomorphic to $D\varphi^{-1}(\underline{0})(V)$. An illustration is given in Figure 2.4.1. Observe that in case M is an open subset of $\mathbb{R}^k$, T_xM coincides with $\mathbb{R}^k$. In general, when taking an element $(x,\delta y)$ of T_xM we only give the component $\delta y \in \mathbb{R}^k$.

Very useful is the next lemma.

<u>Lemma 2.4.5</u>. The Cartesian product $M_1 \times M_2$ of two submanifolds is a submanifold and $\dim(M_1 \times M_2) = \dim(M_1)+\dim(M_2)$. Furthermore, $T_{(x_1,x_2)}(M_1 \times M_2) = \{(x_1,x_2,\delta x_1,\delta x_2) \mid \delta x_1 \in T_{x_1}M_1,\ \delta x_2 \in T_{x_2}M_2\}$.

Now we are ready to define the derivative of a map $f : M_1 \to M_2$ where M_i is a smooth submanifold of $\mathbb{R}^{k_i}$, $i \in \{1,2\}$. The derivative Df_x is a linear map from T_xM_1 to $T_{f(x)}M_2$, and henceforth the best linear approximation to f at x. More formally we get the following definition.

<u>Definition 2.4.6</u>. Let $M_i \subset \mathbb{R}^{k_i}$ be smooth m_i-submanifolds, $i \in \{1,2\}$. Further, $f : M_1 \to M_2$ is a map. Let $x \in M_1$ and let (U_1,φ_1) be the submanifold chart for M_1 at x whereas (U_2,φ_2) is the submanifold chart for M_2 at $f(x)$. The sets V_1 and V_2 are defined as in Definition 2.4.2.

i) The map $f : M_1 \to M_2$ is *smooth* at x if the map $\varphi_2 \circ f \circ \varphi_1^{-1} : V_1 \to V_2$ is smooth at $\underline{0} \in V_1$.

ii) Similarly, f is a C^i-*map* at x, $i \geq 1$, if $\varphi_2 \circ f \circ \varphi_1^{-1} : V_1 \to V_2$ is a C^i-map at $\underline{0} \in V_1$.

iii) Given that f is C^1, the map $T_xf : T_xM_1 \to T_{f(x)}M_2$ is defined by $T_xf(x,\delta x) = (f(x),Df(x).\delta x)$, with

$$Df(x) = D\varphi_2^{-1}(\underline{0}) \circ D(\varphi_2 \circ f \circ \varphi_1^{-1})(\underline{0}) \circ D\varphi_1(x).$$

In fact, $T_x f$ is the derivative of f at x. In general we only give its second component or shortly $Df(x)$. At first sight the definition above seems rather complex. Some clarification might be given by the illustration in Figure 2.4.2. From Figure 2.4.2.a we derive that the map f coincides with the map $\varphi_2^{-1} \circ (\varphi_2 \circ f \circ \varphi_1^{-1}) \circ \varphi_1$ considered on U_1 restricted to M_1. However, unlike f, the functions φ_2^{-1}, $\varphi_2 \circ f \circ \varphi_1^{-1}$ and φ_1 are defined on open sets and their derivatives can be calculated using the standard definition. Using the chain rule we get Definition 2.4.6 (see Figure 2.4.2.b). In fact, Df_x is the derivative of $\varphi_2^{-1} \circ (\varphi_2 \circ f \circ \varphi_1^{-1}) \circ \varphi_1$ restricted to an m-dimensional subspace of $\mathbb{R}^{k_1}$. In case M_1, M_2 are open sets Definition 2.4.6 is equivalent to the standard derivative.

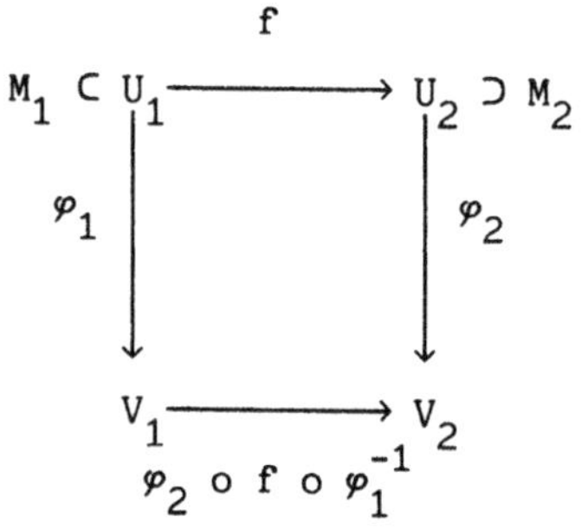

Figure 2.4.2.a

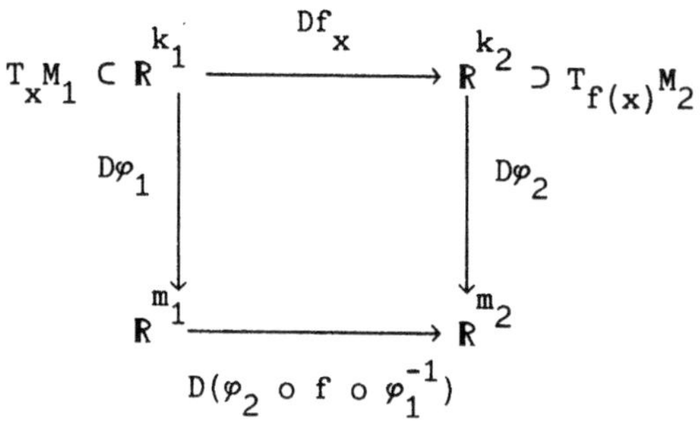

Figure 2.4.2.b

A further illustration is given in Figure 2.4.1. Consider the function $f : \bar{S}^1 \to \bar{S}^1$ defined by $f(x) = -x$. Now the map $T_x f : T_x\bar{S}^1 \to T_{-x}\bar{S}^1$ is defined by $T_x f(x,\delta x) = (-x,-\delta x)$. For short, $Df(x) = -E$ for all $x \in \bar{S}^1$.

Of importance is the rank of the matrix $Df(x)$. Related to the rank are the concepts of regular value and regular point.

Definition 2.4.7. Consider the smooth map $f : M \to N$, with M and N an m- and n-submanifold, respectively. A point $x \in M$ at which $\text{rank}(Df(x)) < \min\{m,n\}$ $(= \min\{m,n\})$ is called a *critical (regular) point* of f. A point $y \in N$ is a *regular value* of f if all points $x \in f^{\leftarrow}(y)$ are regular.

It is well-known from Sard's theorem that the set of regular values of f is dense in N. The importance of regularity stems from the next lemma.

Lemma 2.4.8. If $f : M \to N$ is a smooth map between submanifolds $M \subset \mathbb{R}^k$ and $N \subset \mathbb{R}^\ell$ of dimension m and n, $m \geq n$, and if $y \in N$ is a regular value of f, then the set $f^{\leftarrow}(y) \subset M$ is a smooth submanifold of dimension m-n.

Example 2.4.2. Consider $f : \mathbb{R}^m \to \mathbb{R}$ defined by $f(x) = (\Sigma_{i=1}^m x_i^2)^{\frac{1}{2}}$. Any value $y \neq 0$ is regular. The set $f^{\leftarrow}(y)$, $y \neq 0$, is the smooth (m-1)-submanifold $\{x \in \mathbb{R}^m | \|x\| = y\}$.

Till thusfar we only considered submanifolds without boundary. The definition of a submanifold with boundary is a straightforward extension of Definition 2.4.2.

Definition 2.4.9. A subset $M \subset \mathbb{R}^k$ is a *smooth submanifold with boundary* of $\mathbb{R}^k$ of dimension m if for every point $x \in M$ there exists an open neighbourhood U of x and a smooth function $\varphi : U \to \mathbb{R}^k$ such that $\varphi(x) = \underline{0}$, $\text{rank}(D\varphi(y)) = k$ for all $y \in U$, and $\varphi^{-1}(\bar{V}) = U \cap M$, where $\bar{V} := \{(r_1,\dots,r_k) \in \mathbb{R}^k | r_m \geq 0, r_{m+1} = \dots = r_k = 0\}$. The *boundary* of M, bd(M), is the set of points in M which corresponds to $\text{bd}(\bar{V}) = \{r \in \bar{V} | r_m = 0\}$. The *interior* of M is the set $M \backslash \text{bd}(M)$.

Example 2.4.3. Consider the disk $D^m := \{x \in \mathbb{R}^m | \|x\| \leq 1\}$. This set is a smooth m-submanifold with boundary. The boundary of D^m is the set $\{x \in \mathbb{R}^m | \|x\| = 1\}$.

It is easy to verify that the boundary of an m-submanifold with boundary has dimension m-1 whereas the interior is of dimension m. We now extend Lemma 2.4.8 to submanifolds with boundary.

Lemma 2.4.10. Consider the smooth map $f : M \to N$ from an m-submanifold of $\mathbb{R}^k$ with boundary to an n-submanifold of $\mathbb{R}^\ell$, $m \geq n$. If $y \in N$ is a regular value both for f and for the restriction $f|_{\text{bd}(M)}$, then $f^{\leftarrow}(y) \subset M$ is a smooth (m-n)-submanifold with boundary. Furthermore, $bd(f^{\leftarrow}(y)) = f^{\leftarrow}(y) \cap bd(M)$.

Example 2.4.4. Take the map of Example 2.4.2 but now defined on $V_1^m := \{(x_1,\ldots,x_m)^T \in \mathbb{R}_+^m | x_i > 0,\ i \in \{2,\ldots,m\}\}$. Then $f^{\leftarrow}(y)$, $y \neq 0$, is the smooth (m-1)-submanifold $\{x \in V_1^m | \|x\| = y\}$, with boundary $\{x \in V_1^m | \|x\| = y$ and $x_1 = 0\}$ of dimension m-2.

The Lemmas 2.4.8 and 2.4.10 deal with the preimage of a point. But what about the preimage of a submanifold? For this generalization we need the notion of transversality.

Definition 2.4.11. Let $f : M \to N$ be a smooth map with $M \subset \mathbb{R}^k$ and $N \subset \mathbb{R}^\ell$ smooth submanifolds of dimension m and n. The map f is *transversal* to a submanifold Z of N, denoted $f \pitchfork Z$, if for all $x \in f^{\leftarrow}(Z)$

$$Df(x)(T_xM) + T_yZ = T_yN, \text{ where } y = f(x).$$

Thus, $f \pitchfork Z$ if for all $x \in f^{\leftarrow}(Z)$ we have that for every $\delta y \in T_{f(x)}N$ there are $\delta z \in T_{f(x)}Z$ and $\delta x \in T_xM$ such that $\delta y = \delta z + Df(x).\delta x$. If $f^{\leftarrow}(Z) = \emptyset$ then this condition always holds. Furthermore, the notion of transversality is indeed a generalization of regularity. The two coincide if Z consists of one single point. In case $Z = \{y\}$, $T_{f(x)}Z = \{\underline{0}\}$, and $f \pitchfork \{y\}$ means that for all $x \in f^{\leftarrow}(y)$ we have that for every $\delta y \in T_{f(x)}N$ there are $\delta x \in T_xM$ such that $\delta y = Df(x).\delta x$. In other words, $Df(x)$ must be of full rank.

Example 2.4.5. We consider the maps $f_a : \mathbb{R}^1 \to \mathbb{R}^2$ defined by $f_a(x) = (x, x^2-a)$, $a \in \mathbb{R}$. Let Z be the x-axis in $\mathbb{R}^2$. Then $f_a \pitchfork Z$ if and only if (iff) $a \neq 0$. For illustration see Figure 2.4.3.

Lemma 2.4.12. Let $M \subset \mathbb{R}^k$ be a submanifold with boundary and let $N \subset \mathbb{R}^\ell$ be a boundaryless submanifold. If the smooth maps $f : M \to N$ and $f|_{bd(M)} : bd(M) \to N$ are transversal to a boundaryless submanifold $Z \subset N$, then the preimage $f^{\leftarrow}(Z)$ is a submanifold with boundary of M. Moreover, $\dim(M)-\dim(f^{\leftarrow}(Z)) = \dim(N)-\dim(Z)$ and $bd(f^{\leftarrow}(Z)) = f^{\leftarrow}(Z) \cap bd(M)$. Furthermore, given $x \in f^{\leftarrow}(Z)$ the tangent space to $f^{\leftarrow}(Z)$ consists of all $\delta x \in T_xM$ satisfying $Df(x).\delta x \in T_{f(x)}Z$.

The expression for $\dim(f^{\leftarrow}(Z))$ is often formulated in terms of its codimension. The codimension of a submanifold X in a surrounding manifold Y is defined as $\mathrm{codim}(X) = \dim(Y)-\dim(X)$. In Lemma 2.4.12 we then get $\mathrm{codim}(f^{\leftarrow}(Z)) = \mathrm{codim}(Z)$.

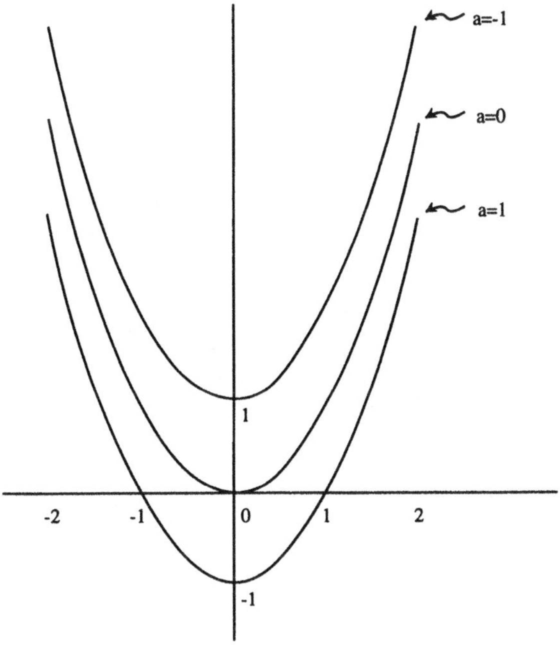

Figure 2.4.3. For $a \neq 0$, the map f_a is transversal to the x-axis in $\mathbb{R}^2$.

The last notion we consider here is that of a homotopy.

Definition 2.4.13. Let $N \subset \mathbb{R}^{\ell}$ and $M \subset \mathbb{R}^k$ be submanifolds and let $M\times[0,1]$ denote the submanifold of $\mathbb{R}^{k+1}$ consisting of all (x,t) with $x \in M$ and $t \in [0,1]$. Two mappings $f,g : M \to N$ are called *smoothly homotopic* if there exists a smooth map $F : M\times[0,1] \to N$ with $F(x,0) = f(x)$ and $F(x,1) = g(x)$, $x \in M$. The map F is called a *smooth homotopy* between f and g.

Example 2.4.6. Consider the mappings $f,g : \mathbb{R} \to \mathbb{R}$ defined by $f(x) = x^2$ and $g(x) = x$. These mappings are smoothly homotopic. As a homotopy can serve the map $F : \mathbb{R}\times[0,1] \to \mathbb{R}$ defined by $F(x,t) = tx + (1-t)x^2$.

CHAPTER 3

Existence of adjustment processes

In this chapter we investigate under what conditions the processes to be considered in this monograph exist and converge. For that we consider one specific process. All processes treated in this monograph, except the ones in Chapter 6, have the same properties and their existence can be studied along the same lines.

The process we consider here is a price adjustment process for a pure exchange economy and has been introduced by van der Laan and Talman [1987a]. Starting from an arbitrarily chosen price vector it leads to an equilibrium price vector via a sequence of price adaptations. These price adaptations are induced by the starting vector and the sign pattern of the excess demand vector. First, we derive conditions for existence and convergence directly from its definition. Next, we put the process in a homotopy framework and derive convergence conditions in that setting. The latter approach is done because it is more convenient. Besides, the homotopy approach enables us to state conditions for monotonic convergence.

Solving the equilibrium problem via the homotopy approach means that we start in a solution to a trivial problem on an artificial set. The trivial problem on the artificial set is then continuously deformed to the real problem on the set of interest. By following a path of solutions for these intermediate problems, which starts with the trivial solution, eventually an equilibrium is reached. We show that the path traced by the price adjustment process can be viewed upon as being the projection of such a homotopy path on the price space.

But first we deal with another price adjustment process developed more recently. Not only because of its own value but also because of reasons of exposition. It turns out that this process fits very well in a very simple and transparant homotopy. The treatment of this process thus also serves as an introduction to the other process being more difficult to analyse.

The set-up of this chapter is as follows. Section 3.1 is introductory. There we briefly sketch an exchange economy model and introduce the two adjustment processes. We also give some economic intuition. In Section 3.2 we treat the most simple process. The Sections 3.3 and 3.4 deal with the process of van der Laan and Talman [1987a]. In Section 3.3 we consider its existence right from the definition, whereas in Section 3.4 we use the homotopy approach. This chapter is partly based on Doup, van den Elzen, and Talman [1989].

3.1. Two price adjustment processes for a pure exchange economy

Before discussing the processes we first briefly review the standard model of a pure exchange economy. In such an economy there are, say, n+1 commodities and a finite number of consumers, each having a preference relation and a vector of initial endowments. Exchanges of goods are based on relative prices. All consumers in this economy exchange goods in order to maximize their utility under the budget constraint imposed by their initial wealth. An equilibrium price vector is a vector of prices at which for all goods demand equals supply while no consumer can improve upon his situation. All relevant information of such an economy can be captured in an excess demand function which relates to each price vector the corresponding vector of excess demands. Thus, an excess demand function, denoted by z, can be seen as a function from the set $\mathbb{R}^{n+1}_{++}$ to $\mathbb{R}^{n+1}$.

For $p \in \mathbb{R}^{n+1}_{++}$, if $z_j(p)$, $j \in I_{n+1}$, is positive, then the market for good j is in excess demand at the price vector $p = (p_1,\dots,p_{n+1})^T$ with p_h denoting the price of commodity $h \in I_{n+1}$, whereas this market is in excess supply at p in case $z_j(p) < 0$. Market j is said to be in equilibrium at $p \in \mathbb{R}^{n+1}_{++}$ if $z_j(p) = 0$. An equilibrium price vector is a vector $p^* \in \mathbb{R}^{n+1}_{++}$ at which $z(p^*) = \underline{0}$. Standard conditions on z are

i) $z(\lambda p) = z(p)$, $\forall \lambda > 0$ and $\forall p \in \mathbb{R}^{n+1}_{++}$ (homogeneity)

ii) $p \cdot z(p) = 0$, $\forall p \in \mathbb{R}^{n+1}_{++}$ (complementarity) (3.1.1)

iii) $\forall p \in bd(\mathbb{R}^{n+1}_{+}) \backslash \{\underline{0}\}$ $\exists \delta > 0$ $[p' \in \{x \in \mathbb{R}^{n+1}_{++} \mid \|p-x\| < \delta\}$ and

$$p_j = 0] \Rightarrow z_j(p') > 0 .$$

Furthermore, we assume that z is smooth on $\mathbb{R}^{n+1}_{++}$. To guarantee the latter certain convexity and monotonicity conditions on the preference relations of the consumers are needed.

The economic interpretation of (3.1.1) is straightforward. Condition i) indicates that only relative prices matter. Condition ii) is also known as Walras' law and says that all consumers spend their total income. Condition iii) is a desirability condition, roughly stating that if the price of a good is (relatively) small, the demand for it exceeds its supply. In Kamiya [1984] it is proved that these conditions are sufficient for the existence of an equilibrium price vector. In fact he uses weaker conditions. We need more to prove the existence of the price adjustment process.

The first process we consider is defined on the set of prices given by

$$B_v := \{p \in \mathbb{R}^{n+1}_{++} | p \cdot v = p \cdot p\}, \qquad (3.1.2)$$

where $v \in \mathbb{R}^{n+1}_{++}$ denotes the starting price vector. By restricting ourselves to B_v no information about this economy is lost. This is due to Condition i) on z which allows for normalizing the prices. The set B_v can be seen as a specific normalization dependent on v. When taking a price vector p in $\mathbb{R}^{n+1}_{++}$, then, given v, $p \cdot v = \alpha p \cdot p$ for some number $\alpha \in \mathbb{R}_+$. We normalize p to the vector $q := \alpha p$ which lies in B_v while $z(q) = z(p)$. In this way each price vector p in $\mathbb{R}^{n+1}_{++}$ is related to a unique vector q in B_v with the same excess demands. Now, the process follows a path of price vectors in B_v originating in v. For a price vector $p \neq v$ on the path it holds that

$$z(p) = \beta(p - v), \quad \text{for some } \beta \geqq 0. \qquad (3.1.3)$$

Thus, at each price vector p on the path it holds that the related excess demand vector is a nonnegative multiple of the difference between p and the starting prices. This implies that at markets revealing excess demand (supply) prices have been increased (decreased) from the starting prices, which is economically appealing. As soon as β becomes 0, an equilibrium price vector is reached. Notice the similarity between (3.1.3) and the classical Walrasian adjustment process if we replace the initial price vector by the (in time) previous price vector. In the sequel we refer to this process as the <u>proportional process</u>.

The second process has also a strong resemblance with the Walras' process. It operates on the interior of the n-dimensional unit simplex S^n. Thus, here the price vectors are normalized by dividing each price by the sum of the prices. From the starting vector $v \in \text{int}(S^n)$ the process follows a path of prices p which all satisfy for $j \in I_{n+1}$,

$$\begin{aligned} p_j/v_j &= \max_h p_h/v_h \quad \text{if } z_j(p) > 0 \\ \min_h p_h/v_h \leq p_j/v_j &\leq \max_h p_h/v_h \quad \text{if } z_j(p) = 0 \qquad (3.1.4) \\ p_j/v_j &= \min_h p_h/v_h \quad \text{if } z_j(p) < 0. \end{aligned}$$

Because this process focusses on the sign pattern of the excess demand vector it is called the <u>sign process</u>. The similarity with the Walras' process occurs in particular at v. If we assume that at v no market is in equilibrium, it follows from (3.1.4) that the process leaves v by increasing the prices of goods in excess demand and decreasing those of goods in excess supply. However, the increases and decreases are relative to the starting price vector. As soon as a price vector is generated at which a good previously being in excess supply (demand) becomes in equilibrium then its price is relatively not further decreased (increased) but adapted in order to keep the good in equilibrium. In general, price vectors are generated such that the relative prices of goods in excess demand (supply) are maximal (minimal), whereas the relative prices of goods in equilibrium vary between these bounds. More details can be found in van der Laan and Talman [1987a]. Finally, we remark that the path of this process can be followed arbitrarily close by the simplicial algorithm presented in Doup, van der Laan, and Talman [1987].

3.2. The proportional process

In this section we propose a homotopy related to the proportional adjustment process as defined by (3.1.3). We show that the set of vectors satisfying (3.1.3) can be seen as the projection on B_v of the set of zero points of this homotopy. From this we derive that in general there exists a path of vectors $p \in B_v$ satisfying (3.1.3) and leading from the vector v to a vector $\tilde{p}$ with $\|z(\tilde{p})\|$ arbitrarily close to zero.

Let v again be the starting price vector. The homotopy function h related to the proportional process starting from v is defined by

$$h(p,\delta) = (1-\delta)(v - p) + \delta z(p) , \qquad (p,\delta) \in \mathbb{R}^{n+1}_{++}\times[0,1]. \tag{3.2.1}$$

The function h is smooth because of the smoothness of z. When going from level zero to level one, the function h smoothly deforms the trivial function, $h(p,0) = v - p$, into the function of interest, $h(p,1) = z(p)$. Note that the domains on both levels are the same. That is why we call h a homotopy in standard form. Furthermore, observe that (v,0) is the unique zero point of h on level zero, whereas the set of zero points on level one coincides with the set of equilibrium price vectors. In the sequel we show that there exists a smooth path in $h^{\leftarrow}(\underline{0})$ connecting (v,0) and a point $(\tilde{p},\tilde{\delta})$, with $\|z(\tilde{p})\|$ arbitrarily close to zero. The projection of this path on $\mathbb{R}^{n+1}_{++}$ then gives a path of price vectors from v to $\tilde{p}$ satisfying (3.1.3).

First, we restrict h to the domain $\mathbb{R}^{n+1}_{++}\times[0,\tilde{\delta}]$, with $\tilde{\delta} \in (0,1)$. Clearly, h is then a smooth function from an (n+2)-manifold with boundary into an (n+1)-manifold. The boundary of the domain of h is equal to $(\mathbb{R}^{n+1}_{++}\times\{0\}) \cup (\mathbb{R}^{n+1}_{++}\times\{\tilde{\delta}\})$. So, if $\underline{0} \in \mathbb{R}^{n+1}$ is a regular value of h then $h^{\leftarrow}(\underline{0})$ is a smooth 1-manifold, i.e. a collection of disjoint smooth paths and loops. Moreover, an end point of a path in $h^{\leftarrow}(\underline{0})$ either lies in $\mathbb{R}^{n+1}_{++}\times\{0\}$, i.e. equals (v,0), or lies in $\mathbb{R}^{n+1}_{++}\times\{\tilde{\delta}\}$, i.e. equals $(\tilde{p},\tilde{\delta})$ (cf. Lemma 2.4.10).

In order to show that $h^{\leftarrow}(\underline{0})$ contains a path from (v,0) to a point in $\mathbb{R}^{n+1}_{++}\times\{\tilde{\delta}\}$ first notice that $h(p,\delta) = 0$, $0 \leqq \delta \leqq \tilde{\delta}$, implies $p\cdot(v-p) = 0$, since $p\cdot z(p) = 0$ and $\delta \neq 1$. Hence, all zero points of h lie in $B_v\times[0,\tilde{\delta}]$.

To get a better insight in B_v observe that B_v is the part of the ball in $\mathbb{R}^{n+1}$ around $\frac{1}{2}v$ and passing through v, which lies in the positive orthant, i.e.

$$B_v = \{p \in \mathbb{R}^{n+1}_{++} \mid (p - \tfrac{1}{2}v)\cdot(p - \tfrac{1}{2}v) = \tfrac{1}{4}v\cdot v\}.$$

Since the closure of B_v is a compact subset of $\mathbb{R}^{n+1}_{+}\backslash\{\underline{0}\}$ and because of the smoothness and the desirability condition on z, there exists an ε, $0 < \varepsilon < \min_h v_h$, such that $z_j(p) > 0$ whenever $p_j \leqq \varepsilon$ and $p \in B_v$. Now, let $B_v(\varepsilon)$ be the compact subset of $\mathbb{R}^{n+1}_{++}$ defined by

$$B_v(\varepsilon) = \{p \in B_v \mid p_j \geqq \varepsilon \text{ for all } j \in I_{n+1}\}.$$

The next lemma shows that all zero points of h in $B_v\times[0,\bar{\delta}]$ lie in $\text{int}(B_v(\varepsilon))\times[0,\bar{\delta}]$.

<u>Theorem 3.2.1</u>. Let (p,δ) be a point in $B_v\times[0,\bar{\delta}]$. Then $h_i(p,\delta) > 0$ if $p_i \leqq \varepsilon$.

<u>Proof</u>. By definition we have for all $(p,\delta) \in B_v\times[0,\bar{\delta}]$,

$$h_j(p,\delta) = (1-\delta)(v_j-p_j)+\delta z_j(p) , \qquad j \in I_{n+1}.$$

If $p_j \leqq \varepsilon$, we obtain with $v_j > \varepsilon$ that $v_j-p_j > 0$ whereas from above we know that $z_j(p) > 0$.

□

Consequently, if $\underline{0}$ is a regular value of h, then the path P, having (v,0) as an end point cannot intersect the boundary of $B_v(\varepsilon)$ between the levels 0 and $\bar{\delta}$. Thus, P must have another end point on $B_v(\varepsilon)\times\{0\}$ or $B_v(\varepsilon)\times\{\bar{\delta}\}$. Since (v,0) is the only zero point of h on level 0 and h is transversal on $B_v(\varepsilon)\times\{0\}$, the other end point is a point $(\tilde{p},\bar{\delta})$ on level $\bar{\delta}$. Moreover, all other paths in $h^{\leftarrow}(\underline{0})$, if any, also lie in $B_v(\varepsilon)\times[0,\bar{\delta}]$ and connect two zero points of h on level $\bar{\delta}$, whereas all loops in $h^{\leftarrow}(\underline{0})$ lie in $B_v(\varepsilon)\times(0,\bar{\delta})$. The latter follows from Lemma 2.4.10. There it is stated that $\text{bd}(h^{\leftarrow}(\underline{0})) = h^{\leftarrow}(\underline{0}) \cap \text{bd}(\mathbb{R}^{n+1}_{++}\times[0,\bar{\delta}])$. Thus, $h^{\leftarrow}(\underline{0})$ cannot touch $\text{bd}(\mathbb{R}^{n+1}_{++}\times[0,\bar{\delta}])$. An illustration of $h^{\leftarrow}(\underline{0})$ is given in Figure 3.2.1 in case

$n = 1$. In that figure, the set $h^{\leftarrow}(\underline{0})$ consists of the path P and a path Q connecting two vectors on level $\tilde{\delta}$. Notice that $h^{\leftarrow}(\underline{0})$ cannot contain loops when $n = 1$. This because (p,δ_1) and (p,δ_2) in $h^{\leftarrow}(\underline{0})$ imply $\delta_1 = \delta_2$. Furthermore, from the figure it is obvious that B_v is indeed a normalization of the price space.

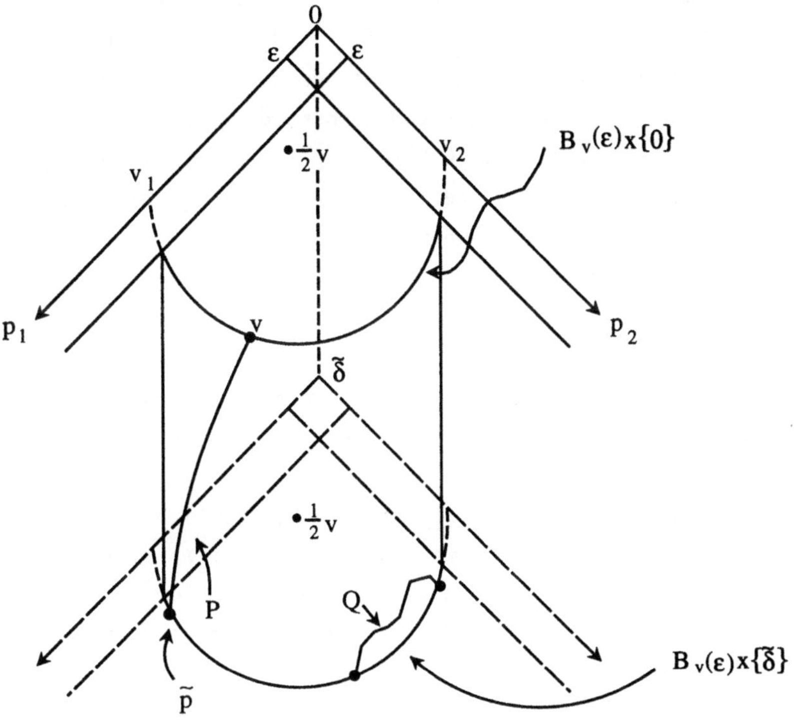

Figure 3.2.1. Illustration of $h^{\leftarrow}(\underline{0})$ in $B_v(\varepsilon)\times[0,\tilde{\delta}]$, $n = 1$.

For $\underline{0}$ to be a regular value of h, it is required that the derivative of h is of full rank at all zero points of h. Given a point (p,δ) in $h^{\leftarrow}(\underline{0})$ the derivative is denoted as the $((n+1)\times(n+2))$-matrix $Dh(p,\delta)$. The first n+1 columns, $D_p h$, contain the derivatives of h with respect to p while the last column, $D_\delta h$, is the derivative of h with respect to δ. More precisely, the Jacobian matrix at (p,δ) is equal to

$$Dh(p,\delta) = [-(1-\delta)E + \delta Dz(p) \mid p - v + z(p)],$$

where $Dz(p)$ is the $((n+1)\times(n+1))$-matrix of derivatives of z at p. Thus, the value $\underline{0}$ is regular for h if the matrix $Dh(p,\delta)$ is of rank $n+1$ for all $(p,\delta) \in B_v(\varepsilon)\times[0,\bar{\delta}]$ for which $h(p,\delta) = 0$.

Thusfar we have given conditions guaranteeing the existence of a smooth path P in $h^{\leftarrow}(\underline{0})$ connecting $(v,0)$ and a point $(\tilde{p},\bar{\delta})$. We still have to show that the projection P' of P on $\mathbb{R}^{n+1}_{++}$ gives the path of the proportional process. First observe that, assuming $z(v) \neq \underline{0}$, P' must be a smooth curve in $B_v(\varepsilon)$ connecting v and a point $\tilde{p}$. This follows from the fact that $(p,\delta_1) \in h^{\leftarrow}(\underline{0})$ and $(p,\delta_2) \in h^{\leftarrow}(\underline{0})$ imply $\delta_1 = \delta_2$. More precisely, the derivative $D_\delta h(p,\delta) = p - v + z(p)$ has rank 1 along the path P. Observe that this only does not hold when at $p = v$ we have $z(p) = \underline{0}$. But then P is the line segment $\{v\}\times[0,\bar{\delta}]$ and $P' = \{v\}$. So, if v is not a zero point of z, the path P' is a smooth curve in $B_v(\varepsilon)$. From (3.2.1) it is obvious that at any point p, $p \neq v$, along P', we have that $z(p) = \beta(p - v)$ for some $\beta \gneq 0$. Thus, indeed P' coincides with the path of points followed by the proportional adjustment process. An illustration for $n = 2$ is given in Figure 3.2.2.

Note further that if $h(p,\delta) = 0$, $(p,\delta) \neq (v,0)$, and so $z(p) = \beta(p - v)$ for some $\beta \geq 0$, we must have $\beta = (1-\delta)/\delta$. Thus, β decreases if δ increases and therefore δ gives an indication for how close the components of $z(p)$ are to zero. We say that the process converges monotone if δ increases monotone from 0 to $\bar{\delta}$ along the path P. The latter only holds if $D_p h(p,\delta) = -(1-\delta)E + \delta Dz(p)$ always has rank $n+1$ along P. Furthermore, by chosing $\bar{\delta}$ arbitrarily close to 1, the norm of $z(\tilde{p})$, with $(\tilde{p},\bar{\delta}) \in h^{\leftarrow}(\underline{0})$, can be made arbitrarily close to zero. The latter because $z(\tilde{p}) = (1-\bar{\delta})(\tilde{p} - v)/\bar{\delta}$.

Instead of the natural normalization on B_v one could also a priori normalize the price vectors to lie in the set T defined by $T = \{\hat{p} \in \mathbb{R}^n_{++} | \Sigma^n_{i=1}\hat{p}_i < 1\}$. After deleting the last component of z, one can then define a homotopy in standard form on $T\times[0,1]$ with respect to $\tilde{z}$, where for $j \in I_n$, $\tilde{z}_j(\hat{p}) = z_j((\hat{p}^\top,1-\Sigma^n_{i=1}\hat{p}_i)^\top)$ (see Kamiya [1990]). This homotopy, however, does not take into account $z_{n+1}(p)$ at all and therefore differs from the one introduced above.

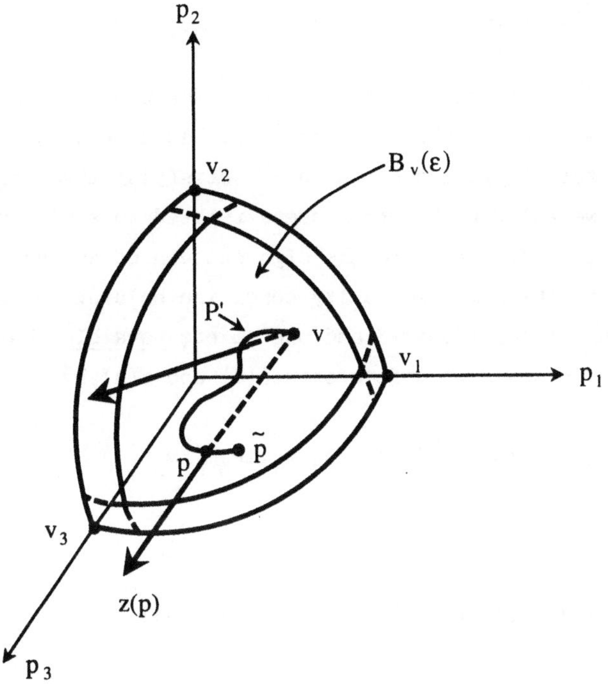

Figure 3.2.2. Illustration of P' for n = 2.

3.3. Direct approach towards the sign process

In this section we consider the convergence and existence of the sign process. We give conditions under which certain subsets of vectors satisfying (3.1.4) form smooth paths. Further, transversality and non-degeneracy assumptions are needed to guarantee that these subsets can be linked to a piecewise smooth path.

First, we pick an arbitrary vector in $\text{int}(S^n)$ to be the starting vector, again denoted by v. Furthermore, from the conditions on z we derive the existence of an η, $0 < \eta < (n+1)^{-1}$, such that for all p in $\text{int}(S^n)$ and all $j \in I_{n+1}$, $z_j(p) > 0$ if $p_j \leqq \eta$. Thus, equilibrium vectors cannot lie very near the boundary of S^n. Since $p_j < \eta v_j$ implies $p_j/v_j < \max_h p_h/v_h$ and $z_j(p) > 0$, it follows from (3.1.4) that price vectors generated by the sign process always lie in the set $S^n(\eta)$ defined by

$$S^n(\eta) = \{p \in S^n | p_j \geqq \eta v_j,\ j \in I_{n+1}\}.$$

In fact, $S^n(\eta)$ is also dependent on v. Throughout this section we surpress this dependency in the notation. Now, let us return to (3.1.4) and denote by D the set of price vectors in S^n satisfying these conditions. For our purposes we split up D into subsets related to a sign vector. Here, such a sign vector corresponds to the sign pattern of an excess demand vector. Because of the complementarity condition holding for z, only sign vectors containing at least one +1 and one -1 are <u>feasible</u>. Related to each feasible sign vector s we define sets A(s) and Y(s) by

$$A(s) = \{p \in S^n(\eta) | p_j/v_j = \min_h p_h/v_h \qquad \text{if } s_j = -1$$

$$p_j/v_j = \max_h p_h/v_h \qquad \text{if } s_j = +1\},$$

and

$$Y(s) = \text{cl}(\{p \in S^n(\eta) | \text{sgn}(z(p)) = s\}).$$

Observe that the definition of A(s) corresponds to the left part of (3.1.4), whereas the definition of Y(s) corresponds to the right part. From this it is obvious that for each p in D there is a sign vector s such that p lies in $D(s) := A(s) \cap Y(s)$. In other words $D = \cup_s D(s)$, where the union is over all feasible sign vectors. The sets A(s) are such that they subdivide $S^n(\eta)$. An illustration of this subdivision is given in Figure 3.3.1. Here, the vertices of $S^n(\eta)$ are denoted by $e^\eta(j)$, $j \in I_{n+1}$. More precisely, $e_i^\eta(j) = \eta v_i$ if $i \neq j$ and $e_i^\eta(j) = 1-(1-v_j)\eta$ if $i = j$. The dimension of A(s) equals $1+|I^0(s)|$.

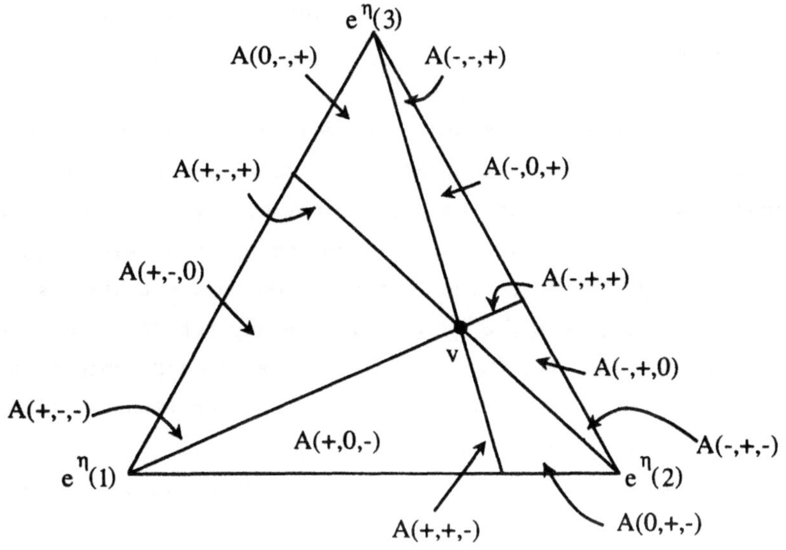

Figure 3.3.1. Subdivision of $S^n(\eta)$ into subsets $A(s)$, $n = 2$. The set $A((-1,+1,0)^T)$ is denoted by $A(-,+,0)$, etc.

Below, we give rather general conditions such that for given sign vector s the set $D(s)$ is a smooth 1-manifold, i.e. it consists of a disjoint union of smooth paths and loops. The idea is the following. We consider the function z_s, being the restriction of z to a smooth manifold with boundary $\tilde{A}(s)$, where $\tilde{A}(s)$ is such that $D(s) = \tilde{A}(s) \cap Y(s)$. We take $\tilde{A}(s)$ instead of $A(s)$ because the latter set is a manifold with corners which prevents the applicability of the transversality Lemma 2.4.12 we want to use. Later on we show how to construct $\tilde{A}(s)$, for the moment it suffices to know that $\dim(\tilde{A}(s)) = \dim(A(s)) = |I^0(s)|+1$. Observe that $D(s)$ equals the closure of the set $\overleftarrow{z_s}(0_s)$ with $0_s \subset \mathbb{R}^{n+1}$ defined by $0_s = \{x \in \mathbb{R}^{n+1} | \operatorname{sgn}(x) = s\}$. Now, if $z_s \pitchfork 0_s$ then $\operatorname{codim}(D(s)) = \operatorname{codim}(0_s) = |I^0(s)|$ and therefore $D(s)$ is a smooth 1-manifold with boundary (see Lemma 2.4.12). Thus, the transversality condition mentioned is sufficient for the existence of $D(s)$.

Let us now express this condition in terms of the function z. According to Definition 2.4.11 we have that $z_s \pitchfork 0_s$ if for all vectors $\alpha \in \mathbb{R}^{n+1}$ there are vectors $\gamma \in 0_s$ and $\bar{p} \in T_p\tilde{A}(s)$, $p \in D(s)$, such that $Dz(p)\bar{p} + \gamma = \alpha$. We note that this is an open condition, i.e. when it holds for some function z it also holds for an open environment around z in the Whitney topology.

To construct $\tilde{A}(s)$ from A(s) the following two observations are crucial. Firstly, we assume that at v, $\mathrm{sgn}(z_j(v)) \neq 0$, $\forall j \in I_{n+1}$. This is a very weak restriction. From that it is obvious that there exists an environment around v consisting of price vectors p at which $\mathrm{sgn}(z_j(p)) = \mathrm{sgn}(z_j(v)) \neq 0$, $\forall j$. Thus, no such vector p lies in a set D(s), $s \neq s^0 = \mathrm{sgn}(z(v))$. Secondly, no $p \in S^n \backslash S^n(\eta)$ lies in any set D(s). Because of all of this we can "smooth" the faces of a set A(s) and obtain a smooth manifold with boundary, $\tilde{A}(s)$, with the required conditions, i.e. $\dim(\tilde{A}(s)) = \dim(A(s))$, $\tilde{A}(s) \subset \mathrm{int}(S^n)$, and $D(s) = \tilde{A}(s) \cap Y(s)$.

The path of prices from v to an equilibrium satisfying (3.1.4) exists of a sequence of adjacent subsets D(s), for different sign vectors s. When going from one D(s) to another, say $D(\bar{s})$, we want only one of the following two cases to happen. For some unique index j, either, $\bar{s}_j = 0$, $s_j \neq 0$, and $\bar{s}_i = s_i$ for $i \neq j$, or $\bar{s}_j \neq 0$, $s_j = 0$, and $\bar{s}_i = s_i$ for $i \neq j$. In these cases when going from D(s) to $D(\bar{s})$ the dimension of the set A(s) increases (decreases) with one whereas the dimension of Y(s) decreases (increases) with one, making that the dimension of D(s) remains one. Let us consider the first case. Suppose the process starts at v with $s^0 = \mathrm{sgn}(z(v))$. Because s^0 does not contain zeros, $D(s^0)$ consists of (possibly more than one) closed line segments lying in the 1-manifold $A(s^0)$. One end point of a line segment is v, the other end point of that segment is a vector p at which $z_j(p)$ is zero for some j. We assume that this occurs for a unique index. Concerning the other case, consider for example the set $D((-1,0,+1)^T)$. A path in this set may contain an end point in the interior of $A((-1,+1,+1)^T)$. From that point on we can continue with a path in the set $D((-1,+1,+1)^T)$ in $A((-1,+1,+1)^T)$. What we need here is that $D((-1,0,+1)^T)$ intersects $A((-1,+1,+1)^T)$ transversally in its interior. For example, in Figure 3.3.1 it is not allowed that $D((-1,0,+1)^T)$ touches $A((-1,+1,+1)^T)$ or intersects this set in its boundary.

Let us summarize the foregoing. We have given conditions on z guaranteeing that each nonempty D(s) is a smooth 1-manifold with boundary. At boundary points of D(s) lying in bd(A(s)) we need to assume that D(s) intersects that boundary transversally in the interior of a facet of $A(s)$ and that the sign pattern of z is still equal to s. At boundary points with changing sign of the function z we need to assume a condition guaranteeing that just one sign component of z changes. Only at the equilibrium two or more components of z may become zero at the same time. The last assumption is that $z_j(v) \neq 0$, $\forall j \in I_{n+1}$. If all of this holds then there exists a piecewise smooth path of price vectors p satisfying (3.1.4) connecting v and an equilibrium.

Finally, we briefly mention another approach for proving the existence of processes. It uses the concept of primal-dual manifolds as introduced by Kojima and Yamamoto [1982]. An application to our processes is given in van den Elzen, van der Laan, and Talman [1985]. However, essentially that method is equivalent to the one presented here.

3.4. Homotopy approach towards the sign process

In this section we state conditions under which the sign process converges towards an equilibrium by putting this process in a homotopy framework. Here we will not derive explicit conditions as done in Section 3.2. This would only bring a lot of formulas. We rather confine ourselves to stating the conditions in general terms such as regularity of certain values. The conditions are more complicated than in Section 3.2 because we here have to use a homotopy that is not in standard form. The latter means that the domains on the two levels are different.

The homotopy function h suited for this process is defined on the convex hull V of $U\times\{0\}$ and $S^n(\eta)\times\{1\}$, where U is given by

$$U = \mathrm{conv}(\{v + e(i) - e(j) \mid i,j \in I_{n+1},\ i \neq j\}).$$

Again $h(x,\delta)$, $(x,\delta) \in V$, will be a convex combination of two function values, defined on U and $S^n(\eta)$ respectively. To get a proper definition of h we relate each (x,δ) in V to a unique pair (u,p), with $u \in U$ and $p \in S^n(\eta)$. To define (u,p) we first subdivide V in an appropriate way into cells. This subdivision is determined by the subdivision of $S^n(\eta)$ on level one into subsets $A(s)$. The cell $C(\emptyset)$ is given by

$$C(\emptyset) = \text{conv}(U\times\{0\},\{v\}\times\{1\}),$$

i.e. $C(\emptyset)$ is equal to the set $\{(x,\delta) \in \mathbb{R}^{n+1}\times[0,1] \mid \Sigma_{j=1}^{n+1}x_j = 1,\ -(1-\delta) \leq x_j-v_j \leq 1-\delta,\ j \in I_{n+1}\}$. Furthermore, for a feasible sign vector s let the face U(s) of U be given by

$$U(s) = \text{conv}(\{v + e(i) - e(j) \mid s_i = +1,\ s_j = -1,\ i,j \in I_{n+1}\}).$$

It is easy to show that U(s) is equal to the set $\{u \in \mathbb{R}^{n+1} \mid u = v + \Sigma_{h=1}^{n+1}\mu_h s_h e(h),\ \mu_h \geq 0$ for $h \in I_{n+1}$, and $\Sigma_{s_h=+1}\mu_h = \Sigma_{s_h=-1}\mu_h = 1\}$. Clearly, U(s) is of dimension $n-|I^0(s)|-1$. The set V is now subdivided into $C(\emptyset)$ and cells C(s), s being a feasible sign vector, where C(s) is the convex hull of $U(s)\times\{0\}$ and $A(s)\times\{1\}$. Observe that the cells C(s) all have dimension n+1 and that also $C(\emptyset)$ is a cell of dimension n+1. These cells are illustrated in Figure 3.4.1.

In Theorem 3.4.2 we prove that $C(\emptyset)$ together with all the sets C(s) indeed form a subdivision of V. More precisely, we show that for any $(x,\delta) \in V$ with $0 < \delta < 1$, there either exists a unique feasible sign vector s such that for unique vectors p in A(s) and u in int(U(s))

$$x = (1-\delta)u + \delta p, \tag{3.4.1}$$

or (3.4.1) holds with $p = v$ and $u \in \text{int}(U)$ uniquely defined by $u = (1-\delta)^{-1}(x - \delta v)$. We call p and u the projection of the point (x,δ) on level 1 and level 0, respectively.

Now, we are ready to define an appropriate homotopy function on V. This function is given by

$$h(x,\delta) = (1-\delta)(v - u) + \delta\tilde{z}(p),$$

where $\tilde{z}_j(p) = p_j z_j(p)$, $j \in I_{n+1}$, and u and p are given by (3.4.1). Because u and p are uniquely determined by (x,δ), the function h is well-defined on V. Moreover, h is a piecewise smooth function from V to $\mathbb{R}^{n+1}$ deforming the trivial function f with $f(u) = v - u$ on level 0 into the function $\tilde{z}$ on

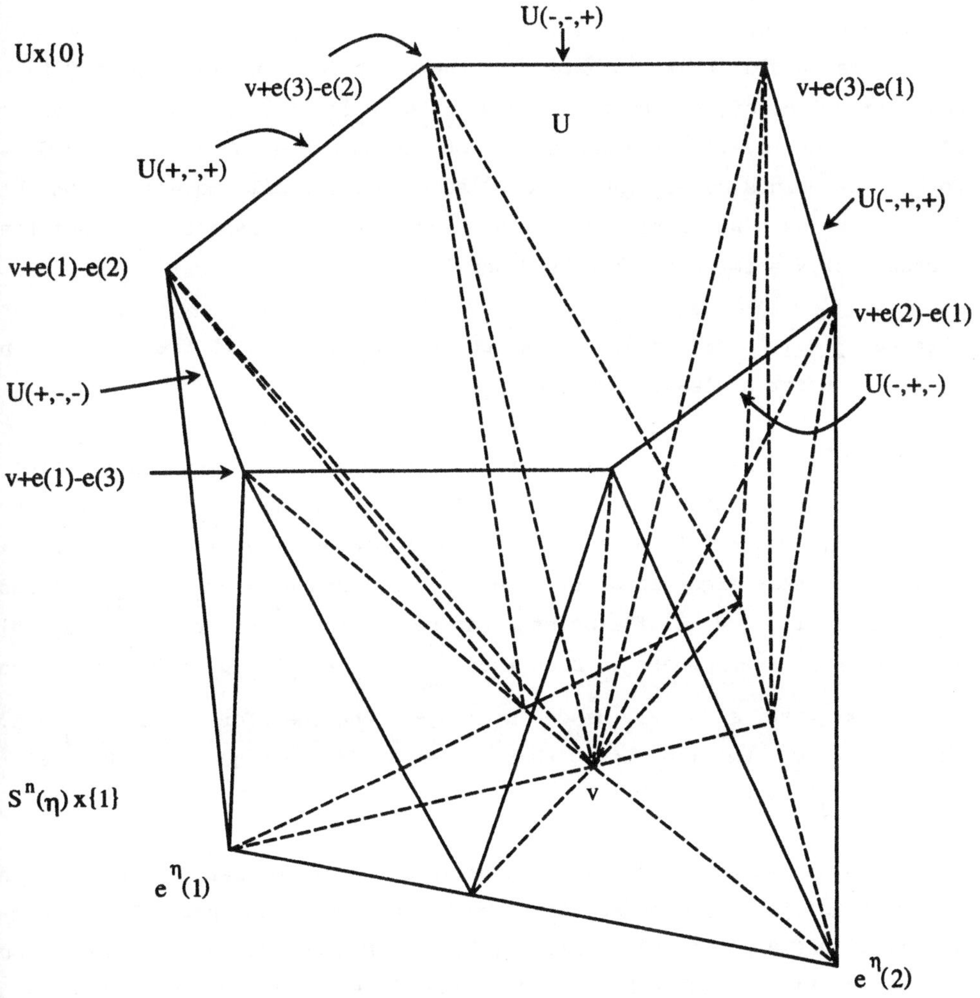

Figure 3.4.1. Subdivision of V into the cells C(∅) and C(s), n = 2. The set $U((-1,+1,+1)^T)$ is denoted by U(-,+,+), etc.

level 1. In fact, h is smooth on each cell C(s) and also on C(∅). Again, we are interested in the set $h^{\leftarrow}(\underline{0})$, i.e. the set of points (x,δ) in V such that for some feasible sign vector s, $x = (1-\delta)u + \delta p$ with $p \in A(s)$ and $u = v + \Sigma_h \mu_h s_h e(h) \in \text{int}(U(s))$ (or $p = v$ and $u \in \text{int}(U)$), while $\tilde{z}(p) = (1-\delta)\delta^{-1}\Sigma_h \mu_h s_h e(h)$. Hence, if $p \neq v$, we have $z_j(p) = (1-\delta)\delta^{-1}\mu_j s_j/p_j$, $j \in$

I_{n+1}, i.e. p satisfies (3.1.4). Observe that $z(p) = \underline{0}$ when $\delta = 1$ and that $x = v$ when $\delta = 0$.

We would like to have that the set $\overleftarrow{h}(\underline{0})$ contains a path P in V connecting the unique zero point (v,0) of h on level 0 with a zero point p^* of z on level 1. The projection of the path P on $S^n(\eta)$ then yields a path P' of points connecting v and p^* such that for all points p on P' (3.1.4) holds. We first prove that a path in $\overleftarrow{h}(\underline{0})$ cannot intersect the boundary of V between the levels 0 and 1.

Theorem 3.4.1. Let (x,δ) be a point in bd(V) with $\delta \notin \{0,1\}$. Then $h_j(x,\delta) > 0$ for at least some $j \in I_{n+1}$.

Proof. It is easy to see that the boundary of V consists of $U\times\{0\}$, $S^n(\eta)\times\{1\}$, and $\cup_s(\text{conv}(U(s)\times\{0\}, \{p \in A(s)|p_j = \eta v_j \text{ if } s_j = -1\}\times\{1\}))$. So, let (x,δ), $0 < \delta < 1$, be a point in the boundary of V, i.e. $x = (1-\delta)u + \delta p$ with for some sign vector s, u in int(U(s)) and $p_j = \eta v_j$ if $s_j = -1$ and $p_j \geq \eta v_j$ if $s_j \neq -1$. The point u can be written as $u = v + \Sigma_{h=1}^{n+1}\mu_h s_h e(h)$ for some positive numbers μ_h such that $\Sigma_{s_h=+1}\mu_h = \Sigma_{s_h=-1}\mu_h = 1$. Let j be an index with $s_j = -1$. Hence, $u_j = v_j-\mu_j$ and therefore $v_j-u_j = \mu_j > 0$. Since $p_j = \eta v_j$, we also have that $z_j(p) > 0$. Consequently, $h_j(x,\delta) = (1-\delta)(v_j-u_j)+\delta p_j z_j(p) > 0$.
□

Next, we show under what conditions the set of zero points of h indeed consists of piecewise smooth paths and loops, one path, P, connecting (v,0) and a point $(p^*,1)$ with $z(p^*) = \underline{0}$. Notice that h is linear on $C(\emptyset)$ and smooth on each (n+1)-dimensional set C(s). Furthermore, h is a function from the (n+1)-dimensional set V to the n-manifold 0^n defined by

$$0^n = \{q \in \mathbb{R}^{n+1}|\Sigma_{i=1}^{n+1}q_i = 0\}.$$

Hence, $\overleftarrow{h}(\underline{0}) \cap C(\emptyset)$ is a line segment connecting (v,0) with a point in the boundary of $C(\emptyset)$. Furthermore, if $\underline{0}$ is a regular value of h restricted to C(s), then $\overleftarrow{h}(\underline{0}) \cap C(s)$ is a smooth 1-manifold, i.e. a collection of disjoint smooth paths and loops. Technically, the proof is somewhat more complex. Because C(s) is a manifold with corners we have to construct a set $\tilde{C}(s)$ such that $\overleftarrow{h}(\underline{0}) \cap \tilde{C}(s) = \overleftarrow{h}(\underline{0}) \cap C(s)$ while $\tilde{C}(s)$ is a smooth manifold with boundary. $\tilde{C}(s)$ can be constructed in the same way as $\tilde{A}(s)$ in Section 3.3. An end point of a path in $C(s)$ lies in the boundary of $C(s)$. We have to assume that a path in $\overleftarrow{h}(\underline{0}) \cap C(s)$ intersects the boundary transversally in the interior of a facet of $C(s)$. The latter guarantees that an end point of a path in $\overleftarrow{h}(\underline{0}) \cap C(s)$ either lies in $bd(V)$ or is also an end point of a path in $\overleftarrow{h}(\underline{0}) \cap C(\emptyset)$ or in $\overleftarrow{h}(\underline{0}) \cap C(\bar{s})$ for some unique $\bar{s} \neq s$. More precisely, in the latter case, since a facet of $C(s)$ not in the boundary of V is equal to either $C(\emptyset) \cap C(s)$ or to $conv(U(\hat{s})\times\{0\},A(s)\times\{1\})$, with $\hat{s}_i = 0$ and $s_i \neq 0$ for some unique $i \in I_{n+1}$ while $\hat{s}_j = s_j$ for $j \neq i$, or to $conv(U(s)\times\{0\},A(\tilde{s})\times\{1\})$, with $\tilde{s}_i \neq 0$ and $s_i = 0$ for some unique $i \in I_{n+1}$ while $\tilde{s}_j = s_j$ for $j \neq i$, the vector $\bar{s}$ is equal to either $\hat{s}$ or $\tilde{s}$. Linking the paths in $\overleftarrow{h}(\underline{0}) \cap C(\emptyset)$ and in $\overleftarrow{h}(\underline{0}) \cap C(s)$ for different s in this way, we obtain that if $\underline{0}$ is a regular value of h on each $C(s)$ and transversality on the boundaries is assumed then $\overleftarrow{h}(\underline{0})$ consists of piecewise smooth loops and paths. One path, P, has $(v,0)$ as end point on level 0 in $C(\emptyset)$. Since an end point of a path cannot lie in $bd(V)$ between levels 0 and 1, the other end point of P lies on level one and induces a zero point of z. The path P is linear on $C(\emptyset)$ and smooth on each $C(s)$ it intersects. As argued above, the projection P' of P on $S^n(\eta)\times\{1\}$ yields the path of points of the adjustment process induced by (3.1.4). Note that in $C(\emptyset)$, $\overleftarrow{h}(\underline{0})$ is a line segment connecting $(v,0)$ and a point in $bd(C(s^0))$.

We already derived that a point $(x,\delta) \in \overleftarrow{h}(\underline{0}) \cap C(s)$ is related to a price vector $p \in S^n(\eta)$ for which $z_j(p) = (1-\delta)\delta^{-1}\mu_j s_j/p_j$, with $\mu_j > 0$, when $s_j \neq 0$. Thus, this process converges monotonically to an equilibrium iff $(1-\delta)\delta^{-1}\mu_j/p_j$ decreases monotonically for all $j \in I_{n+1}$ from $\delta = 0$ to $\delta = 1$.

We conclude the description of the homotopy h on V by proving the following theorem.

<u>Theorem 3.4.2</u>. Given $(x,\delta) \in V$, $0 < \delta < 1$, there either exist a unique feasible sign vector s and vectors $p \in A(s)$ and $u \in int(U(s))$ such that $x = (1-\delta)u + \delta p$, or there exists a unique $u \in int(U)$ such that $x = (1-\delta)u + \delta v$, i.e. $(x,\delta) \in int(C(\emptyset))$.

<u>Proof</u>. First we verify when $(x,\delta) \in \mathrm{int}(C(\emptyset))$. This is clearly the case if and only if $-(1-\delta) < x_j - v_j < (1-\delta)$ for all $j \in I_{n+1}$. The corresponding $u \in \mathrm{int}(U)$ and p are given by $u = (1-\delta)^{-1}(x - \delta v)$ and $p = v$.

In the remaining of the proof we show that when $(x,\delta) \notin \mathrm{int}(C(\emptyset))$ we can find an s as stated in the theorem. First we rank the numbers $(x_j-(1-\delta)v_j)/\delta v_j$, $j \in I_{n+1}$, in increasing order. Without loss of generality we may assume that

$$(x_1-(1-\delta)v_1)/\delta v_1 \leqq (x_2-(1-\delta)v_2)/\delta v_2 \leqq \ldots \leqq (x_{n+1}-(1-\delta)v_{n+1})/\delta v_{n+1}.$$

In the sequel we often replace the expression $(x_j-(1-\delta)v_j)/\delta v_j$ by a_j, $j \in I_{n+1}$. From the definition of C(s) we derive that if $(x,\delta) \in C(s)$ then $x_j = (1-\delta)(v_j+\mu_j s_j)+\delta p_j$, $j \in I_{n+1}$, with $p \in A(s)$ and the μ_j's defined as before. Thus, $a_j := p_j/v_j+(1-\delta)\mu_j s_j/\delta v_j$, $j \in I_{n+1}$. Combined with the definition of A(s) we get that for all $j \in I_{n+1}$,

$$a_j < p_j/v_j = \min_h p_h/v_h \qquad \text{if } s_j = -1$$

$$a_j = p_j/v_j \qquad \text{if } s_j = 0$$

$$a_j > p_j/v_j = \max_h p_h/v_h \qquad \text{if } s_j = +1.$$

Since $a_1 \leqq \ldots \leqq a_{n+1}$, the sign vector s must be such that there are two indices $k, \ell \in I_{n+1}$ with $k < \ell$ such that $s_j = -1$ if $j \leqq k$, $s_j = 0$ if $k < j < \ell$, and $s_j = +1$ if $j \geqq \ell$.

After this first observation we determine the index k and the value f of $\min_h p_h/v_h$. The values f and k have to be such that $\Sigma_{j=1}^k \mu_j = 1$ and $a_k < f \leqq a_{k+1}$. We find f by gradually increasing $\min_h p_h/v_h$ from a_1 and therefore increasing $\Sigma_{j=1}^k \mu_j$, subsequently for k = 1,2,..., from 0. Suppose that we cannot find such an f. Then we must meet the situation that $p_j/v_j = \min_h p_h/v_h$ for $j \in \{1,\ldots,n\}$ and $a_{n+1} = \min_h p_h/v_h = p_{n+1}/v_{n+1}$. Therefore $p = v$ and $a_{n+1} = 1$. We argue that in this situation $x_j - v_j < 1-\delta$ for all $j \in I_{n+1}$ because $x_h - v_h \geqq 1-\delta$ for some $h \in I_{n+1}$ implies $a_h = (x_h-(1-\delta)v_h)/\delta v_h \geqq 1 + (1-\delta)/\delta v_h > 1$. But this is in contradiction with $1 = a_{n+1} \geqq a_j$, $j \in \{1,\ldots,n\}$. Thus, we obtain that we cannot find an appropriate $\min_h p_h/v_h$ if and only if $x_j - v_j < 1-\delta$ for all $j \in I_{n+1}$. Similarly, we can search for the value of $\max_h p_h/v_h$ by decreasing the maximum from a_{n+1}. We then get that this procedure does not succeed if and only if $x_j - v_j > -(1-\delta)$, $j \in I_{n+1}$.

Since we consider the case that $(x,\delta) \notin \mathrm{int}(C(\emptyset))$, there must exist an index $j \in I_{n+1}$ such that $x_j - v_j \geqq 1-\delta$ or $x_j - v_j \leqq -(1-\delta)$. Let us consider the first case. From above we obtain that we can find an index k and a value f of $\min_h p_h/v_h$ such that $a_k < f \leqq a_{k+1}$. We now show that we can always find an appropriate value g of $\max_h p_h/v_h$ by increasing the maximum from f. Indeed, g must be such that for some $\ell > k$, $\Sigma_{j=\ell}^{n+1} \mu_j = 1$ and $a_{\ell-1} \leqq g < a_\ell$. Because this sum decreases if we increase $\max_h p_h/v_h$ and equals zero if $\max_h p_h/v_h = a_{n+1}$, it suffices to show that $\Sigma_{j=\ell}^{n+1} \mu_j \geqq 1$ when $\ell = k+1$ and $\max_h p_h/v_h = f$. Since $\Sigma_{j=1}^{k} \mu_j = 1$, we obtain in case $\max_h p_h/v_h = f$, and $\ell = k+1$,

$$\Sigma_{j=k+1}^{n+1} \mu_j = \Sigma_{j=k+1}^{n+1} (a_j - f)\delta v_j/(1-\delta) = \Sigma_{j=k+1}^{n+1} (x_j - (1-\delta)v_j - \delta f v_j)/(1-\delta)$$

$$= (1-(1-\delta)-\delta f)/(1-\delta) + \Sigma_{j=1}^{k} (-x_j + (1-\delta)v_j + \delta f v_j)/(1-\delta)$$

$$= \delta(1-f)/(1-\delta) + \Sigma_{j=1}^{k} \mu_j = \delta(1-f)/(1-\delta) + 1 \,.$$

Now suppose $\Sigma_{j=k+1}^{n+1} \mu_j < 1$. But then $f > 1$. Because $f = \min_h p_h/v_h$ this is in contradiction with the fact that both p and v lie in $S^n(\eta)$.

Along the same lines we can treat the case in which $x_j - v_j \leqq -(1-\delta)$ for some $j \in I_{n+1}$. Then we know that there exists a suitable maximum and from that we can show the existence of an appropriate minimum.

Thus, in the foregoing we proved, given an $(x,\delta) \notin \mathrm{int}(C(\emptyset))$, the existence of a feasible sign vector s such that $(x,\delta) \in C(s)$. Besides, we found a related $p \in A(s)$. From the construction above it follows that the point $u := (x - \delta p)(1-\delta)^{-1}$ lies in $\mathrm{int}(U(s))$. Consequently, $x = (1-\delta)u + \delta p$ with $p \in A(s)$ and $u \in \mathrm{int}(U(s))$ uniquely determined.

□

The theorem implies that the collection of cells $C(s)$, s a feasible sign vector in $\mathbb{R}^{n+1}$, together with $C(\emptyset)$ form a subdivision of V into cells. The intersection of $C(s)$ with $S^n(\eta)\times\{1\}$ is equal to $A(s)\times\{1\}$ and the intersection of $C(s)$ with $U\times\{0\}$ is equal to $U(s)\times\{0\}$. The intersection of $C(\emptyset)$ with $S^n(\eta)\times\{1\}$ is equal to $\{(v,1)\}$ and the intersection of $C(\emptyset)$ with $U\times\{0\}$ is equal to $U\times\{0\}$. These sets have been illustrated in Figure 3.4.1. More generally, we can consider on level δ, $0 \leq \delta \leq 1$, sets $C_\delta(s) := \{x \in \mathbb{R}^{n+1} | (x,\delta) \in C(s)\}$ and $C_\delta(\emptyset) := \{x \in \mathbb{R}^{n+1} | (x,\delta) \in C(\emptyset)\}$. The union of $C_\delta(\emptyset)$ and the cells $C_\delta(s)$, s a feasible sign vector, equals the set $V_\delta := \{x \in \mathbb{R}^{n+1} | (x,\delta) \in V\}$. In Figure 3.4.2 the subdivision of V_δ into $C_\delta(\emptyset)$ and $C_\delta(s)$'s is illustrated for three different δ's.

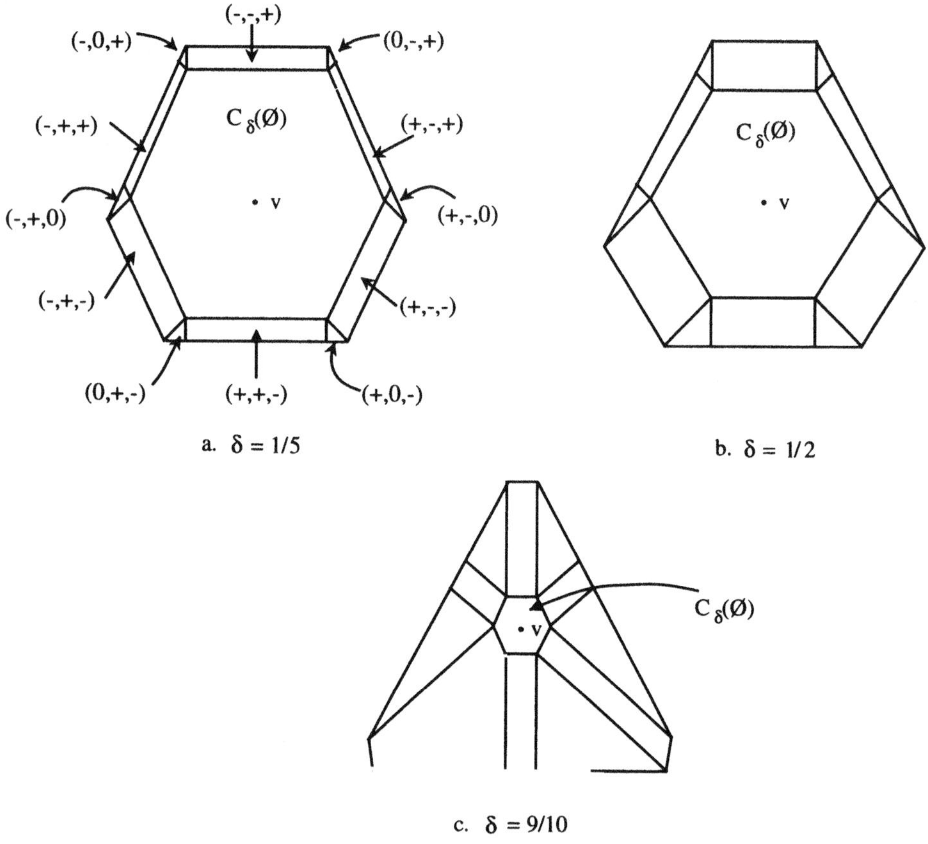

<u>Figure 3.4.2</u>. Subdivision of V_δ into subsets $C_\delta(s)$ and $C_\delta(\emptyset)$ for $n = 2$, $v = (1/4,1/4,1/2)^T$, and $\delta = 1/5$, $1/2$, and $9/10$. The set $C_\delta((-1,-1,+1)^T)$ is indicated by $(-,-,+)$, etc.

CHAPTER 4

An adjustment process for an international trade model

In this chapter we consider price adjustments resulting from shocks in a two country model. For that purpose we present a price adjustment process that reaches an equilibrium price vector when starting from an arbitrarily chosen price vector. The latter vector can be interpreted as the equilibrium price prevailing in the economy before the shock took place.

The international trade model we consider in this paper is very simple. It is a two country model in which both countries supply goods for the domestic market as well as for the international markets (see Mansur and Whalley [1982]). In fact Mansur and Whalley consider a similar model with an arbitrary number of countries. We only consider two countries because all the features of our process can be clarified in this setting. An equilibrium price vector is a vector of prices at which the demand for both the domestic goods and the goods traded on the common market equal their respective supplies. In other words, an equilibrium vector is a zero point of an appropriate excess demand function.

For applying our process we first have to rewrite the model. Here we follow the lines of van der Laan [1985] who reformulated the model by making use of its specific structure. This reformulation introduces exchange rates between the domestic currencies and an international currency, and a balance of payments for each country. Thus, application of our process to the model not only guarantees convergence, but it also makes it possible to consider movements in balances of payments and exchange rates. In fact this makes our process very interesting from an economic viewpoint. Finally, we remark that the path of the process can be followed arbitrarily close by the exponent-ray algorithm presented in Doup, van den Elzen, and Talman [1987].

This chapter consists of five sections. In Section 4.1 we present the basic model and its reformulation. In Section 4.2 we treat a somewhat artificial example to get more insight into the problem. Section 4.3 contains a formal description of the price adjustment process. In Section 4.4 we give an economic interpretation of that process. Finally, in Section 4.5 we present a simple numerical example. This chapter is based on van den Elzen and van der Laan [1986, 1989].

4.1. The model

As already mentioned, our model deals with two countries. Concerning the goods there are domestic goods being produced and traded within a country, and internationally traded or common goods. The model describes a pure exchange economy. Initially, each agent in the economy possesses a bundle of goods. They maximize their utilities by exchanging goods. Exchanges take place at equilibrium prices, i.e. prices at which the demand for each good is equal to its supply.

Let us explain the model more formally. The number of domestic goods in country h, $h \in \{1,2\}$, equals n_h, and these goods are indexed by (h,k), where h denotes the country and $k \in \{1,\dots,n_h\}$ the good. The number of common goods is n_0+1 and they are indexed $(0,k)$, $k \in \{0,1,\dots,n_0\}$. Thus, the total number of goods equals N+1, with $N := n_0+n_1+n_2$. A price vector p is an (N+1)-vector with nonnegative components and can be written as $p = (p_0^\top,p_1^\top,p_2^\top)^\top$ with $p_0 \in \mathbb{R}_+^{n_0+1}$ and $p_h \in \mathbb{R}_+^{n_h}$, $h \in \{1,2\}$. A component p_{hk} of p denotes the price of good k on market h, $h \in \{0,1,2\}$, where the common market is regarded as market 0. More precisely, the price space equals $\bar{R}^{N+1}$, where $\bar{R}^{N+1}$ is the subset of $\mathbb{R}_+^{N+1}\backslash\{\underline{0}\}$, such that for all $p \in \bar{R}^{N+1}$, $(p_0^\top,p_h^\top)^\top$ has at least one positive element for $h \in \{1,2\}$. The excess demand function z^h of country h, $h \in \{1,2\}$, is a continuous function from $\bar{R}^{N+1}$ into $\mathbb{R}^{N+1}$. More precisely, $z^h(p) = (z_0^h(p)^\top,z_1^h(p)^\top,z_2^h(p)^\top)^\top$, where $z_0^h(p) \in \mathbb{R}^{n_0+1}$ denotes the total excess demand of country h for common goods whereas $z_k^h(p) \in \mathbb{R}^{n_k}$ is the total excess demand of the consumers in country h for the domestic commodities of country k, with $k,h \in \{1,2\}$. It is assumed that for both h = 1, and 2, $z_k^h(p) = 0$ if $k \notin \{h,0\}$. Because the excess demands in a country only depend on the domestic and the common goods prices we can write $z^h(p)$ as

$z^h(p_0,p_h)$. Concerning the excess demand z^h of country h we make the following standard assumptions on excess demand functions:

$$\text{i)} \quad z^h(\lambda p_0,\lambda p_h) = z^h(p_0,p_h),\ \lambda > 0 \text{ (homogeneity)}$$

$$\text{ii)} \quad p\cdot z^h(p) = p_0\cdot z_0^h(p_0,p_h) + p_h\cdot z_h^h(p_0,p_h) = 0 \text{ (Walras' law)} \qquad (4.1.1)$$

iii) positive excess demand for goods with price zero (desirability).

Let $z(p) := z^1(p) + z^2(p)$ be the aggregate excess demand at price vector p. At an equilibrium price vector $\bar{p} = (\bar{p}_0^T,\bar{p}_1^T,\bar{p}_2^T)^T \in \bar{R}^{N+1}$ all markets have to be in equilibrium, i.e.

$$\text{i)} \quad z_0(\bar{p}) = z_0^1(\bar{p}_0,\bar{p}_1) + z_0^2(\bar{p}_0,\bar{p}_2) = \underline{0}$$

$$\qquad\qquad (4.1.2)$$

$$\text{ii)} \quad z_h(\bar{p}) = z_h^h(\bar{p}_0,\bar{p}_h) = \underline{0},\ h \in \{1,2\}.$$

It is well-known that (4.1.1) ensures the existence of an equilibrium price vector. This because the equilibrium problem is equivalent to the zero point problem of z on the normalized price space S^N.

To apply our process we first reformulate the equilibrium problem into a zero point problem on the simplotope $S := S^{n_0}\times S^{n_1}\times S^{n_2}$ (see van der Laan [1985]). Related to each element $q = (q_0,q_1,q_2)$ in S we define for both countries a price vector $\pi^h(q) = (\pi_0^h(q),\pi_h^h(q)) \in \mathbb{R}^{n_0+1}\times\mathbb{R}^{n_h}$, $h \in \{1,2\}$, by

$$\pi_{0k}^h(q) = q_{h0}q_{0k} \quad , (0,k) \in I(0)$$

$$\qquad\qquad (4.1.3)$$

$$\pi_{hk}^h(q) = q_{hk} \quad , (h,k) \in I(h)\setminus\{(h,0)\},$$

where $I(h) := \{(h,0),(h,1),\dots,(h,n_h)\}$, $h = 0,1,2$.
So q_{hk}, $k \neq 0$, is the price of the k-th commodity of country h. The components of q_0 are all multiplied by q_{h0} to get the prices of the international commodities for country h. The vectors $\pi^h(q)$, $h \in \{1,2\}$, are defined

in such a way that there is a one-to-one correspondence between a vector q in S at which q_{10} and q_{20} are both positive, and a ray of price vectors p in $\bar{R}^{N+1}$ with $p_0 \neq \underline{0}$, in the sense that q yields the same excess demand vector through $\pi^h(q)$, $h \in \{1,2\}$, as a price vector p on the corresponding ray. Note that we do not consider vectors q in S at which q_{10} or q_{20} is zero. This because our process does not reach such vectors. Also at a vector q related to an equilibrium in $\bar{R}^{N+1}$, both q_{10} and q_{20} have to be positive. This will become clear later on. We illustrate the relation between S and $\bar{R}^{N+1}$ with an example.

Example 4.1.1. Suppose that $(n_0,n_1,n_2) = (2,2,2)$, i.e. there are three common goods and two domestic goods for each country. Now consider the ray $R := \{y \in \bar{R}^7 | y = \lambda x,\ x = (5,2,3,2,1,14,6)^\top,\ \lambda > 0\}$. To derive the related q-vector we first normalize the sum of the common goods prices to one, i.e. we take $\bar{x} = (1/2,1/5,3/10,1/5,1/10,7/5,3/5)^\top$. Hence, $q_0 = (1/2,1/5,3/10)^\top$. Next we find q_1 by solving the equations $q_{11}/q_{10} = \bar{x}_{11} = 1/5$, $q_{12}/q_{10} = \bar{x}_{12} = 1/10$ and $q_{10}+q_{11}+q_{12} = 1$. In this way we find $q_1 = (10/13,2/13,1/13)^\top$ and similar $q_2 = (1/3,7/15,1/5)^\top$. From this we derive that $\pi^1(q) = (5/13,2/13,3/13,2/13,1/13)^\top$ and $\pi^2(q) = (1/6,1/15,1/10,7/15,1/5)^\top$. Clearly, $z^h(\pi^h(q)) = z^h(\bar{x}_0,\bar{x}_h)$, $h \in \{1,2\}$.

Next, we define $\tilde{z}(q) = (\tilde{z}_0^\top(q),\tilde{z}_1^\top(q),\tilde{z}_2^\top(q))^\top \in \mathbb{R}^{N+1}$ as the total excess demand vector at prices $\pi^h(q)$, $h \in \{1,2\}$, i.e.

$$\tilde{z}_0(q) = z_0^1(\pi^1(q)) + z_0^2(\pi^2(q))$$

$$\tilde{z}_h(q) = z_h^h(\pi^h(q)),\ h \in \{1,2\}.$$

Observe that q consists of $\Sigma_{h=0}^2(n_h+1)$ components while $\tilde{z}(q)$ consists of $1+\Sigma_{h=0}^2 n_h$ components. We now construct a function $\bar{z}$ relating to each q a vector with the same number of components. For $q \in S$, we define $\bar{z}(q) = (\bar{z}_0^\top(q),\bar{z}_1^\top(q),\bar{z}_2^\top(q))^\top \in \mathbb{R}^{N+3}$ by

$$\bar{z}_{0k}(q) = \tilde{z}_{0k}(q) \qquad ,\ (0,k) \in I(0)$$

$$\bar{z}_{h0}(q) = \Sigma_{k=0}^{n_0} q_{0k} z_{0k}^h(\pi^h(q)) \qquad ,\ h \in \{1,2\} \qquad (4.1.4)$$

$$\bar{z}_{hk}(q) = \tilde{z}_{hk}(q) \qquad , k \in \{1,\ldots,n_h\},\ h \in \{1,2\}.$$

Observe that for $h \in \{1,2\}$, and for all $q \in S$, (4.1.1.ii) implies

$$q_h \cdot \bar{z}_h(q) = \Sigma_{k=0}^{n_0} \pi_{0k}^h(q) z_{0k}^h(\pi^h(q)) + \Sigma_{k=1}^{n_h} \pi_{hk}^h(q) z_{hk}^h(\pi^h(q)) = 0. \tag{4.1.5a}$$

However, this complementarity condition does not hold for $h = 0$. Then we get

$$q_0 \cdot \bar{z}_0(q) = \bar{z}_{10}(q) + \bar{z}_{20}(q). \tag{4.1.5b}$$

It follows straightforward from (4.1.1.i) and (4.1.1.iii) that $\bar{z}(q^*) = 0$ iff $(q_0^*, p_1^*, p_2^*) \in S^{n_0} \times \mathbb{R}_+^{n_1} \times \mathbb{R}_+^{n_2}$, with $p_{hk}^* = q_{hk}^*/q_{h0}^*$ for $h \in \{1,2\}$ and $k \in \{1,\ldots,n_h\}$, yields an equilibrium. (see also Example 4.1.1).

We remark that the problem of finding a q^* in S with $\bar{z}(q^*) = \underline{0}$ is equivalent with finding a stationary point of $\bar{z}$ on S. That a zero point is a stationary point is clear from (2.2.1). The opposite follows from Theorem 2.2.4. Because $\bar{z}_h$ satisfies the complementarity condition for $h \in \{1,2\}$, whereas also the boundary condition holds, it follows from (2.2.6) that at a stationary point q^* both $\bar{z}_1(q^*)$ and $\bar{z}_2(q^*)$ are equal to zero. But then with (4.1.5b) we must have that $q_0^* \cdot \bar{z}_0(q^*) = 0$. Because of the boundary condition $q_0^* > 0$ and thus also $\bar{z}_0(q^*) = \underline{0}$.

What about the interpretation of $\bar{z}$ and q? From (4.1.3) we see that q_{0k}, $(0,k) \in I(0)$, can be interpreted as the common goods prices denoted in an international currency, e.g. ECU's. The q_{hk}'s for $h \in \{1,2\}$ and $k \neq 0$ denote the prices of the domestic goods in the domestic currency. As we noted already, q_{h0} can be viewed upon as the value of one unit of the international currency in terms of the currency of country h. Thus, an increase of q_{h0} increases the common goods prices in terms of the domestic currency. Moreover, because $\Sigma_{k\neq 0} q_{hk} = 1-q_{h0}$, an increase of q_{h0} also results in a decrease of the sum of the domestic goods prices. With $1-q_{h0}$ as an index of the domestic price level, we can interprete $q_{h0}/(1-q_{h0})$ as the exchange rate. Thus, an increase (decrease) of q_{h0} yields a devaluation (revaluation) of the currency of country h.

Concerning the vector $\bar{z}(q)$, the components $\bar{z}_{0k}(q)$ are the excess demands for the common goods, whereas the $\bar{z}_{hk}(q)$'s for $h \in \{1,2\}$ and $k \neq 0$, denote the excess demands for the domestic goods. From (4.1.4) we see that $\bar{z}_{h0}(q)$ is the value of the excess demand of country h for common goods, expressed in the international currency. In other words, it reveals the situation on the balance of payments. The component $\bar{z}_{h0}(q)$ being positive (negative) indicates that country h faces a deficit (surplus) on its balance.

Given vectors q and $\bar{z}(q)$ we say that $\bar{z}_{hk}(q)$ and q_{hk}, $(h,k) \in I$, correspond to each other. Thus, each excess demand $\bar{z}_{hk}(q)$, $(h,k) \notin \{(1,0),(2,0)\}$, corresponds to a price q_{hk}, whereas the balance of the balance of payments of country $h \in \{1,2\}$, $\bar{z}_{h0}(q)$, corresponds to q_{h0} being the component of q_h determining the exchange rate for country h.

Thus, the reformulation of the model on the simplotope enriches the economic content of the model by the introduction of exchange rates and balances of payments. This enables us to pose questions that could not be dealt with in the basic model. For example, consider the situation in which a certain country has a fixed production capacity represented by its initial endowments. The country has to decide how to divide these endowments between the common market and its home market. In the reformulated model we could take the exchange rate as a target variable and consider the problem how to divide the endowments in order to reach the target. In the sequel of this chapter we discuss a process for the model adjusting prices and exchange rates simultaneously. This process describes how by adaptations of prices and exchange rates a shock in the economy is restored towards a new equilibrium.

4.2. An example

In this section we want to consider an "international trade problem" with only one country, two common goods and one domestic good. Of course, the distinction between common and non-common goods makes no sense in this case and we can solve the equilibrium price vector problem by searching for such a vector of prices on the 2-dimensional unit price simplex S^2. However, here we formulate the problem on the simplotope $S = S^1 \times S^1$ with the two common goods indexed by (0,0) and (0,1) and the non-common good indexed by (1,1). It will appear that this artificial example provides useful insight for attacking the general problem.

Our problem concerns the search for a vector q^* in S at which the "excess demand" $\bar{z}(q^*)$ is equal to zero. Recall from Section 4.1 that $\bar{z}$ is a function from S to $\mathbb{R}^4$ with

$$\bar{z}_{0k}(q) = z_{0k}(\pi(q)) \qquad , k \in \{0,1\}$$

$$\bar{z}_{10}(q) = q_{00}z_{00}(\pi(q)) + q_{01}z_{01}(\pi(q))$$

$$\bar{z}_{11}(q) = z_{11}(\pi(q)),$$

where $\pi(q)$ is the vector of prices $(\pi_{00}(q),\pi_{01}(q),\pi_{11}(q))$ defined by

$$\pi_{0k}(q) = q_{10}q_{0k} \qquad , k \in \{0,1\}$$

$$\pi_{11}(q) = q_{11},$$

and where $z(\pi(q)) = (z_{00}(\pi(q)),z_{01}(\pi(q)),z_{11}(\pi(q)))^{\top}$ is the excess demand at price vector $\pi(q)$.

In this section we do not want to confuse the reasoning with many definitions and formulas. For the moment it suffices to know that the process considers vectors in the intersections of corresponding primal (P) and dual (D) subsets of S. For certain sign vectors $s = (s_{00},s_{01},s_{10},s_{11})^{\top}$ we define a primal set P(s) and a corresponding dual set D(s). Let v be the starting point of the process. We assume that v lies in the interior of S and hence $v_{hk} > 0$ for all (h,k). Now, suppose that some point q lies in the intersection of P(s) and D(s) for certain sign vector s. Then the first set gives information about the position of q with respect to v while the fact that $q \in D(s)$ indicates that the sign pattern of $\bar{z}(q)$

equals s. We can restrict our attention to those sign vectors s for which there exists a q in S with $\text{sgn}(\bar{z}(q)) = s$.

We now consider the particular example. For $\bar{z}_1$ the complementarity condition holds, since

$$q_1 \cdot \bar{z}_1(q) = q_{10}\bar{z}_{10}(q)+q_{11}\bar{z}_{11}(q)$$

$$= q_{10}q_{00}z_{00}(\pi(q))+q_{10}q_{01}z_{01}(\pi(q))+q_{11}z_{11}(\pi(q))$$

$$= \pi_{00}(q)z_{00}(\pi(q))+\pi_{01}(q)z_{01}(\pi(q))+\pi_{11}(q)z_{11}(\pi(q)) = 0.$$

The last step follows from Walras' law. So, for any $q \in S$ we have that $\bar{z}_{10}(q).\bar{z}_{11}(q) \leqq 0$, i.e. the two components of $\bar{z}_1(q)$ are not both positive or negative. With respect to $\bar{z}_0$ the complementarity condition does not hold and hence all sign patterns are possible. On the other hand, it follows from the definition of $\bar{z}_{10}(q)$ that $\bar{z}_{10}(q) \geqq 0$ ($\leqq 0$) if both $\bar{z}_{00}(q)$ and $\bar{z}_{01}(q)$ are nonnegative (nonpositive). If q is in the interior of S and $\bar{z}(q) \neq \underline{0}$ it follows that the collection of possible sign patterns is given by the set $\mathcal{T}$ with

$$\mathcal{T} = \left\{ \begin{bmatrix} + \\ + \\ + \\ - \end{bmatrix}, \begin{bmatrix} - \\ - \\ - \\ + \end{bmatrix}, \begin{bmatrix} + \\ - \\ + \\ - \end{bmatrix}, \begin{bmatrix} + \\ - \\ - \\ + \end{bmatrix}, \begin{bmatrix} - \\ + \\ + \\ - \end{bmatrix}, \begin{bmatrix} - \\ + \\ - \\ + \end{bmatrix}, \begin{bmatrix} + \\ 0 \\ + \\ - \end{bmatrix}, \begin{bmatrix} - \\ 0 \\ - \\ + \end{bmatrix}, \begin{bmatrix} 0 \\ + \\ + \\ - \end{bmatrix}, \begin{bmatrix} 0 \\ - \\ - \\ + \end{bmatrix}, \begin{bmatrix} + \\ - \\ 0 \\ 0 \end{bmatrix}, \begin{bmatrix} - \\ + \\ 0 \\ 0 \end{bmatrix} \right\},$$

where $(+,+,+,-)^T$ denotes the vector $(+1,+1,+1,-1)^T$, etc.

We now define for each $s \in \mathcal{T}$ a primal set P(s) such that the collection of sets P(s) covers $S = S^1 \times S^1$. This covering is given in Figure 4.2.1. The reasoning behind this figure is that prices induced by a vector $q \in P(s)$ are proportionally higher (lower) than in v if the corresponding component of s is positive (negative). Prices corresponding to zero components of s may vary between these relative bounds.

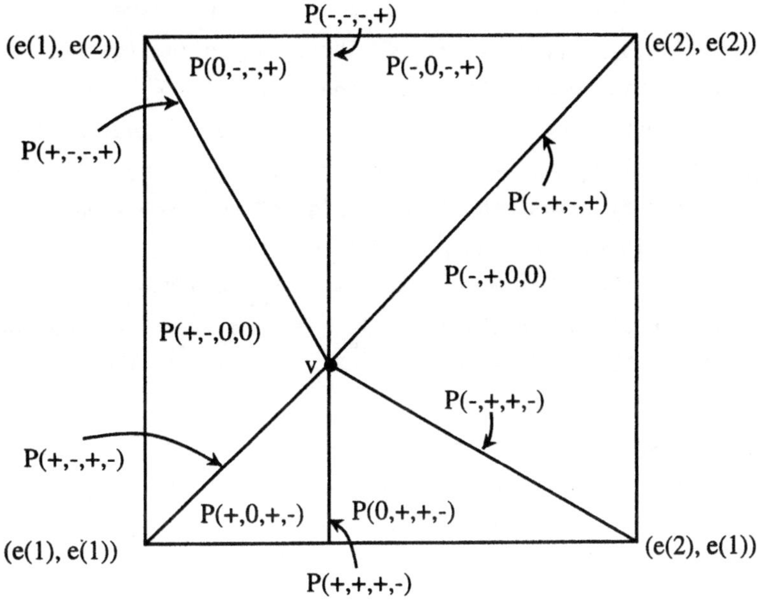

Figure 4.2.1. $S = S^1 \times S^1$. The primal sets $P(s)$, $s \in \mathcal{T}$.

For example consider the primal set $P(s)$ induced by $s = (-1,-1,-1,+1)^T$. For a q in this set we have that $q_0 \leqq v_0$ while $q_{10} < v_{10}$ and $q_{11} > v_{11}$. Hence $\pi_{0k}(q) < \pi_{0k}(v)$, $k \in \{0,1\}$ and $\pi_{11}(q) > \pi_{11}(v)$. Since we are interested in points q in the intersection of corresponding sets $P(s)$ and $D(s)$ this reflects the fact that $\bar{z}_{0k}(q) < 0$, $k \in \{0,1\}$, $\bar{z}_{10}(q) < 0$, and $\bar{z}_{11}(q) > 0$ if $q \in D(s)$. Thus, the prices of the common goods (0,0) and (0,1) should be decreased and the price of the non-common good (1,1) should be increased. Furthermore, q_{10} should be decreased. This reflects a revaluation of the domestic currency of country 1 in order to get rid of the surplus on its balance of payments. Observe that $q_0 \leqq v_0$ implies $q_0 = v_0$ because the sum of the components of both q_0 and v_0 is equal to one. Hence, the price ratios between common goods cannot be adjusted when the common goods are all in excess supply. The same holds when $q_0 \geqq v_0$ if all common goods are in excess demand.

For $q \in P(s)$ with $s = (-1,+1,+1,-1)^T$ we have that $q_{00}/q_{01} < v_{00}/v_{01}$, while $q_{10}/q_{11} > v_{10}/v_{11}$. Thus, $\pi_{00}(q)/\pi_{01}(q)$ is lower than $\pi_{00}(v)/\pi_{01}(v)$. This reflects the fact that for $q \in D(s)$, $\bar{z}_{00}(q) < 0$ and $\bar{z}_{01}(q) > 0$. Furthermore, $q_{10}/q_{11} > v_{10}/v_{11}$ implies that $\pi_{00}(q)+\pi_{01}(q) = q_{10} > v_{10} = \pi_{00}(v)+\pi_{01}(v)$, while $\pi_{11}(q) = q_{11} < v_{11} = \pi_{11}(v)$, which corresponds to the fact that the value $\bar{z}_{10}(q)$ of the excess demands for common goods is positive, whereas $\bar{z}_{11}(q)$ is negative if $q \in D(s)$.

We see that the regions $P(s)$ with $s_{hk} = 0$ for some (h,k) are the convex hull of $P(s')$ and $P(s")$ with $s'_{hk} = -1$ and $s"_{hk} = +1$ whereas $s_{j\ell} = s'_{j\ell} = s"_{j\ell}$ for all $(j,\ell) \neq (h,k)$. For instance $P((-1,0,-1,+1)^T)$ is the convex hull of $P((-1,-1-1,+1)^T)$ and $P((-1,+1,-1,+1)^T)$. Since because of the complementarity condition for $\bar{z}_1$, $\bar{z}_{10}(q) = 0$ iff $\bar{z}_{11}(q) = 0$, we have that $P((-1,+1,0,0)^T)$ is the convex hull of $P((-1,+1,-1,+1)^T)$ and $P((-1,+1,+1,-1)^T)$. The regions $P(s)$ with $s_0 = \underline{0}$ are not defined. This because then $s_{10} = 0$ and so $s = \underline{0}$.

A possible collection of dual sets $D(s)$, $s \in \mathcal{T}$, where $D(s) = \mathrm{cl}(\{q \in S | \mathrm{sgn}(\bar{z}(q)) = s\})$, is illustrated in Figure 4.2.2. The set $D(s)$ with $s_{hk} = 0$ for some (h,k) is the intersection of the two sets $D(s')$ and $D(s")$ with $s'_{hk} = -1$ and $s"_{hk} = +1$ whereas $s_{j\ell} = s'_{j\ell} = s"_{j\ell}$ for all $(j,\ell) \neq (h,k)$. For instance $D((0,+1,+1,-1)^T)$ is the intersection of the sets $D((+1,+1,+1,-1)^T)$ and $D((-1,+1,+1,-1)^T)$, while $D((+1,-1,0,0)^T)$ is the intersection of $D((+1,-1,+1,-1)^T)$ and $D((+1,-1,-1,+1)^T)$.

Figure 4.2.2 shows the "typical" shapes of the sets $D(s)$. For instance consider the point q. Since q_{10} is much smaller than q_{11} and q_{00} almost equals q_{01} we have that both $\pi_{00}(q)$ and $\pi_{01}(q)$ are much smaller than $\pi_{11}(q)$, i.e. the common goods are cheap relatively to the non-common good, so that we may expect that $\bar{z}_{00}(q), \bar{z}_{01}(q) > 0$ and hence that $\bar{z}_{10}(q) > 0$, $\bar{z}_{11}(q) < 0$. Thus, it is very reasonable that q lies in $D((+1,+1,+1,-1)^T)$. Similar arguments hold for other points q in S.

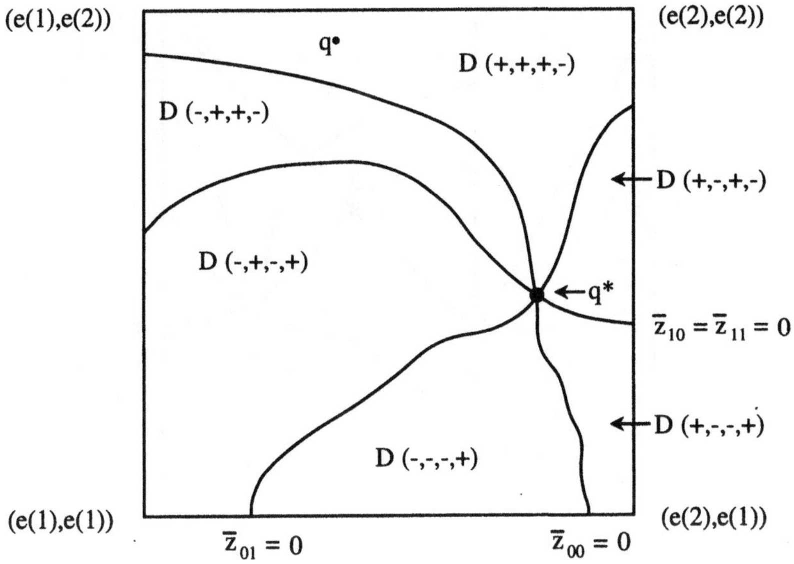

Figure 4.2.2. $S = S^1 \times S^1$. The dual sets $D(s)$, $s \in \mathcal{T}$.

Starting in some point v, the process follows a path of points in the intersections of corresponding primal and dual sets until an equilibrium point q^* is found. This is illustrated in Figure 4.2.3 for the example of Figure 4.2.2 with the starting point as chosen in Figure 4.2.1. Since $v \in D((-1,+1,-1,+1)^T)$ the process leaves v along the ray $P((-1,+1,-1,+1)^T)$ tracing a path of points q in $P((-1,+1,-1,+1)^T) \cap D((-1,+1,-1,+1)^T)$ until the point $\bar{q} \in D((-1,+1,0,0)^T)$ is reached. Then the process continues by tracing a path of points in the intersection of $P((-1,+1,0,0)^T)$ and $D((-1,+1,0,0)^T)$ until the equilibrium point q^* is found.

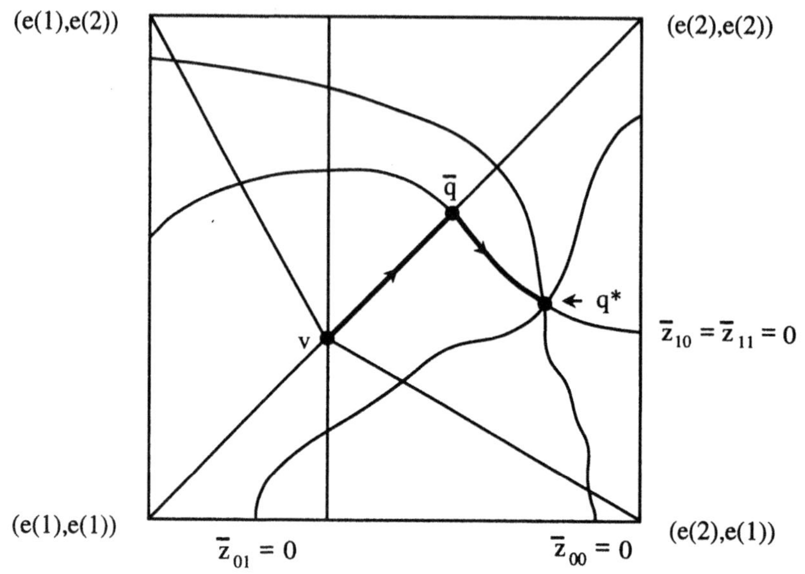

Figure 4.2.3. The path followed by the process, $v \in D((-1,+1,-1,+1)^T)$.

4.3. The process

In this section we present a mathematical description of the price adjustment process. The process can start at any vector v in S and reaches a q^* at which $\bar{z}(q^*) = \underline{0}$ through a path of vectors q in S. At a vector q on the path the process is governed by v and the sign pattern of $\bar{z}(q)$. Very roughly speaking, a component of q is increased (decreased) when the corresponding component of $\bar{z}(q)$ is positive (negative), while if a component of $\bar{z}(q)$ is zero then the corresponding component of q is adjusted such as to keep that component of $\bar{z}(q)$ equal to zero. In the sequel we assume that v lies in the interior of S, i.e. $v_{hk} > 0$ for all $(h,k) \in I$.

The sign pattern of $\bar{z}(q)$ is represented by a sign vector $s = (s_0^T, s_1^T, s_2^T)^T$ in $\mathbb{R}^{N+3}$. Because of (4.1.5), the set $\mathcal{T}$ of feasible sign patterns of $\bar{z}$ is restricted. For example, it is impossible that the components of $\bar{z}(q)$ are all positive. In the sequel we first define $\mathcal{T}$ and related to each element in $\mathcal{T}$ we then define a primal and a dual set, both being subsets of S. A primal set states conditions on the location in S of its elements. It is here that the starting vector plays a major role. The corresponding dual set is induced by the sign pattern of the function values of its elements. The process then only considers vectors of S lying in the intersection of a primal and its corresponding dual set.

Definition 4.3.1. A sign vector s belongs to $\mathcal{T}$ if it satisfies the following conditions:

T1. $s \neq \underline{0}$
T2. $I_h^+(s) = \emptyset$ iff $I_h^-(s) = \emptyset$, $h \in \{1,2\}$
T3. if $s_0 \geq \underline{0}$ $(s_0 \leq \underline{0})$ then $s_{h0} = +1$ $(s_{h0} = -1)$ for at least one h in $\{1,2\}$
T4. if $s_0 = \underline{0}$ then $s_{h0} = 0$, $h \in \{1,2\}$.

The conditions T2 and T3 reflect the properties of $\bar{z}$ as given in (4.1.5). In fact, T2 reflects (4.1.5a) while T3 follows from (4.1.5b). Conditions T1, T2 and T3 imply that there exists an $h \in \{0,1,2\}$ for which $I_h^+(s) \neq \emptyset$ and $I_h^-(s) \neq \emptyset$. Only condition T4 is imposed from outside the model. The reason for this will become clear later on. Observe that we implicitely consider only sign patterns of $\bar{z}(q)$ with $\bar{z}(q) \neq \underline{0}$ and q in the interior of S. For example, if q_1 equals $(0,1,0)^T$ then the corresponding sign pattern of $\bar{z}_1(q)$ equals $(+1,0,+1)$ according to (4.1.1.iii) and (4.1.5a), which contradicts T2. In the sequel we show that our process never reaches a q with some component equal to zero. Confer also the remark made above Example 4.1.1.

Related to each feasible sign vector s we define a primal set P(s).

<u>Definition 4.3.2</u>. Let $s \in \mathcal{T}$ and $q \in S$. Then $q \in P(s)$ if for some b, b_0 and a_h, $h \in \{0,1,2\}$, q satisfies the following conditions, where $0 \leqq b \leqq b_0 \leqq 1$, and $a_h \geqq 1$:

P1. for $h \in \{1,2\}$, and also for $h = 0$ if $s_0 \nleq \underline{0}$,

$$q_{hk} = a_h v_{hk} \qquad \text{if } s_{hk} = +1$$

$$bv_{hk} \leqq q_{hk} \leqq a_h v_{hk} \qquad \text{if } s_{hk} = 0$$

$$bv_{hk} = q_{hk} \qquad \text{if } s_{hk} = -1$$

P2. if $s_0 \leq \underline{0}$,

$$b_0 v_{0k} \leqq q_{0k} \leqq a_0 v_{0k} \qquad \text{if } s_{0k} = 0$$

$$b_0 v_{0k} = q_{0k} \qquad \text{if } s_{0k} = -1.$$

To provide a better insight we will make some remarks. If $s_0 \nleq \underline{0}$ (P1) then all components of a vector q in $P(s)$ related to negative components of s are, relative to v, minimal, i.e. $q_{hk}/v_{hk} = b = \min_I q_{i\ell}/v_{i\ell}$ for all (h,k) with $s_{hk} = -1$. Besides, for each h the components q_{hk} of q with $s_{hk} = +1$ are, relative to v, maximal over all indices in $I(h)$. More precisely, $q_{hk}/v_{hk} = a_h = \max_{I(h)} q_{h\ell}/v_{h\ell}$ if $s_{hk} = +1$. Components of q related to zero components of s are allowed to vary between these bounds, i.e. $q_{hk}/v_{hk} \in [b,a_h]$ if $s_{hk} = 0$. From all of this together it implicitely follows that $q_0 = v_0$ if $s_{0k} = +1$ for all $(0,k) \in I(0)$. If $s_0 \leq \underline{0}$ (P2) then for all q in $P(s)$ it holds that $q_{0k}/v_{0k} = b_0 = \min_{I(0)} q_{0\ell}/v_{0\ell}$ if $s_{0k} = -1$, whereas $q_{hk}/v_{hk} = b$ if $s_{hk} = -1$ and $h = 1$ or 2, where b_0 at least equal to b. In particular, when $s_{0k} = -1$ for all k, we have $q_0 = v_0$ and thus $b_0 = 1$.

Now, consider an $s \in \mathcal{T}$ such that for all h, $s_{hk} = +1$ for just one k, say k_h, while $s_{h\ell} = -1$ for all other $(h,\ell) \in I(h)$. Then the set $P(s)$ is the line segment from v to the vertex $w(s)$ of S with $w_{hk_h}(s) = 1$ and $w_{h\ell}(s) = 0$ for all $\ell \neq k_h$, $h \in \{0,1,2\}$. Observe that along this ray the ratio between prices of commodities with corresponding negative s-component does not change, since all these prices are decreased from v with the uniform factor 1-b. In fact, b decreases from 1 to zero when going from v to w(s). Of course we need a not uniform factor a_h to increase the price of commodity (h,k_h) in order to keep the sum of the components of q_h equal to one. When going from v to w(s), a_h increases from 1 to $1/v_{hk_h}$, $h \in \{0,1,2\}$.

For each sign vector $s \in \mathcal{T}$ with $s_{hk} \neq 0$ for all $(h,k) \in I$, P(s) is a line segment. If $s_0 \not< \underline{0}$ such a ray points to a point $\bar{q}$ of S with $\bar{q}_{hk} = 0$ if $s_{hk} = -1$ and $\bar{q}_{hk} = \bar{a}_h v_{hk}$ if $s_{hk} = +1$, where $\bar{a}_h = (\Sigma_{(h,k)\in I_h^+(s)} v_{hk})^{-1}$. If $s_0 < \underline{0}$ this it still true for $h \in \{1,2\}$ while $\bar{q}_0 = v_0$. Thus, the dimension of P(s) with s such that $s_{hk} \neq 0$ for all $(h,k) \in I$, equals 1. In general, the dimension of P(s) depends on the number of zeros in s. We obtain that

$$\dim(P(s)) = 1 + \Sigma_{h=0}^{2}(|I_h^0(s)|-k_h(s)),$$

where $k_h(s) = 1$ if $|I_h^0(s)| = n_h+1$ and $k_h(s) = 0$ otherwise. The factor $k_h(s)$ stems from the fact that $q_h \in S^{n_h}$, $h \in \{0,1,2\}$, so that q_h is completely determined by n_h of its components.

Some of the primal sets corresponding to different sign vectors coincide. This is obvious for s and $\bar{s} \in \mathcal{T}$ when $s_0 > \underline{0}$, $\bar{s}_0 < \underline{0}$, and $s_h = \bar{s}_h$, $h \in \{1,2\}$. It also holds when $s \in \mathcal{T}$ is such that $s_{0\ell} = +1$ and $s_{0k} = 0$ for all $k \neq \ell$, while for some $h \neq 0$, $s_{h0} = -1$. Then P1 says that for q in P(s) there are b and a_0 with $0 \leqq b \leqq 1 \leqq a_0$ such that $q_{0\ell} = a_0 v_{0\ell}$ and $bv_{0k} \leqq q_{0k} \leqq a_0 v_{0k}$, $k \neq \ell$. Let λ_k be such that $q_{0k} = \lambda_k v_{0k}$ and let $\lambda_r = \min_k \lambda_k$. Then with $b_0 = \lambda_r \geqq b$ we have that q is also in $P(\bar{s})$ with $\bar{s}_{0r} = -1$, $\bar{s}_{0k} = 0$ for all $k \neq r$, and $\bar{s}_h = s_h$, $h \in \{1,2\}$. For example, in Figure 4.3.1 we have that $P((0,-1,+1,-1,-1,+1)^T)$ is equal to $P((+1,0,+1,-1,-1,+1)^T)$ in $S = S^1 \times S^1 \times S^1$.

Observe that the sets P(s) are completely determined by the starting vector v. In Figure 4.3.1 we give an illustration of some sets P(s), $s \in \mathcal{T}$, in case $S = S^1 \times S^1 \times S^1$, i.e. when there are two commonly traded goods and one domestic commodity for each country. Also Figure 4.2.1 may serve as an illustration.

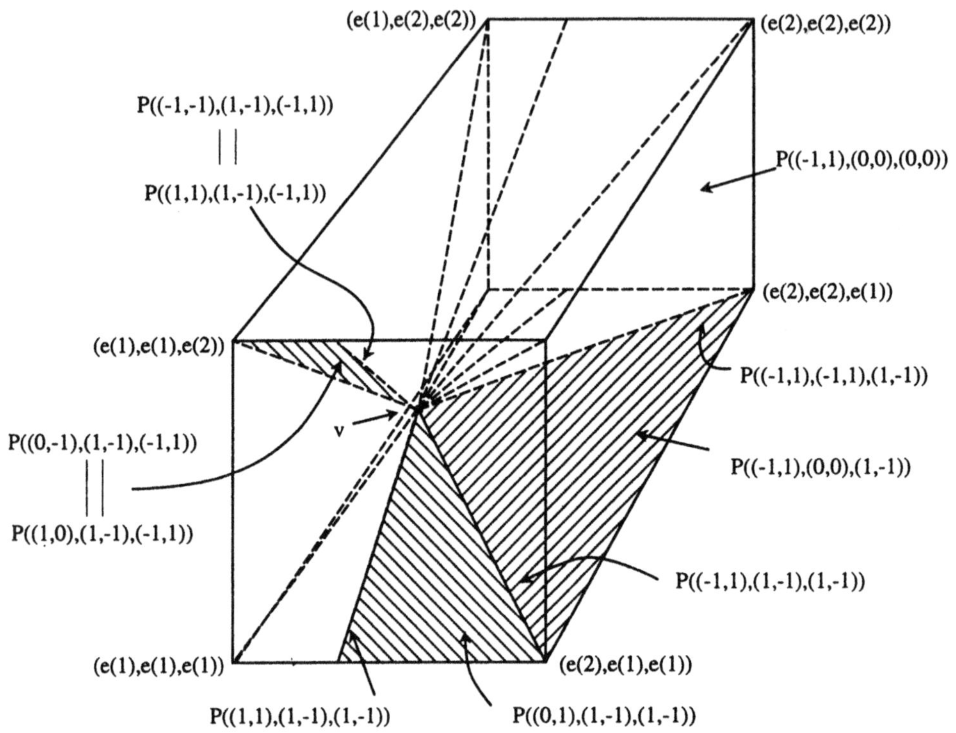

Figure 4.3.1. Illustration of sets P(s) in case $S = S^1 \times S^1 \times S^1$ and $v = ((2/3,1/3)^\top,(5/8,3/8)^\top,(2/3,1/3)^\top)$.

Next we turn to the dual sets which are related to the function $\bar{z}$. For each $s \in \mathcal{T}$ we define the dual subset D(s) of S by

$$D(s) = \mathrm{cl}(\{q \in S \mid \mathrm{sgn}(\bar{z}(q)) = s\}).$$

In the sequel we assume that $\bar{z}$ is a smooth function. What about the dimension of D(s), $s \in \mathcal{T}$? First recall that $\bar{z}$ is a function from S to $\mathbb{R}^{N+3}$ obeying three restrictions, namely

$$q_h \cdot \bar{z}_h(q) = 0 \qquad , h \in \{1,2\}$$

$$q_0 \cdot \bar{z}_0(q) = \bar{z}_{10}(q) + \bar{z}_{20}(q).$$

Because of the first two restrictions we have that for $h \in \{1,2\}$, the (n_h+1)-vector $\bar{z}_h(q)$ is fully determined by n_h of its components. Moreover, by the last restriction, we have that the value of the excess demands for common goods must be equal to the sum of the balances of payments. Hence, we can regard $\bar{z}$ as an unconstrained function $\hat{z}$ from S to the $\Sigma_{h=0}^{2} n_h$-dimensional set $\mathbb{R}^N$. Restricted to D(s), we then may view upon $\bar{z}$ as being a function from D(s) to an $\Sigma_{h=0}^{2}(n_h-|I_h^0(s)|+k_h(s))$-dimensional subset of $\mathbb{R}^N$, denoted $\bar{z}(D(s))$, where $k_h(s)$ is defined as before. The latter subset has a codimension equal to $\Sigma_{h=0}^{2}(|I_h^0(s)|-k_h(s))$. If $\hat{z} : S \to \mathbb{R}^N$ is transversal to $\bar{z}(D(s))$ then D(s) is a smooth manifold in S with a codimension in S equal to $\Sigma_{h=0}^{2}(|I_h^0(s)|-k_h(s))$ (see Lemma 2.4.12). Thus, under fairly general conditions D(s) is a smooth manifold of dimension $\Sigma_{h=0}^{2}(n_h-|I_h^0(s)|+k_h(s))$. Hence, a non-empty intersection $PD(s) := P(s) \cap D(s)$ has dimension equal to $\dim(P(s))+\dim(D(s))-\dim(S) = 1 + \Sigma_{h=0}^{2}(|I_h^0(s)|-k_h(s)) + \Sigma_{h=0}^{2}(n_h-|I_h^0(s)|+k_h(s)) - \Sigma_{h=0}^{2} n_h = 1$.

Thus, for all $s \in \mathcal{T}$ the intersection of P(s) and D(s) not being empty is a 1-dimensional smooth manifold, i.e. it consists of disjoint paths and loops. In the rest of this section we argue that the set $\cup_{s\in\mathcal{T}}$ PD(s) contains under rather general conditions a piecewise smooth path connecting v and a point q^* at which $\bar{z}(q^*) = \underline{0}$. By following this path we get a process of price adjusting. Each smooth piece of the path lies in PD(s) for some specific sign vector s and can be written as the solution curve to some system of differential equations.

For the existence of the path we assume without loss of generality that $s^0 := \mathrm{sgn}(\bar{z}(v))$ contains no zeros. Clearly, $s^0 \in \mathcal{T}$, $v \in PD(s^0)$, $v \in \mathrm{bd}(P(s^0))$, $\dim(P(s^0)) = 1$, and $\dim(D(s^0)) = \dim(S)$. Thus, v must be an end point of a line segment in $PD(s^0)$. Our process starts by following this segment from v. In general, the process follows a curve in some set PD(s), $s \in \mathcal{T}$, being a finite collection of paths and loops. Each path has two end points. In order to guarantee that the process never enters a loop we need that two sets PD(s), PD($\bar{s}$), successively met on the path, intersect transversally in the interior of a facet of P(s) or P($\bar{s}$). Under this condition the process traverses a set PD(s) on the path along a curve connecting two points on the boundary of PD(s). We will show that an end point of a curve in PD(s), $s \in \mathcal{T}$, is either a zero point of $\bar{z}$ or an end point of exactly one path in some PD($\bar{s}$), $\bar{s} \neq s$, $\bar{s} \in \mathcal{T}$. Because of the finite cardinality of $\mathcal{T}$ and the fact that each PD(s) consists of only a finite number of paths this guarantees the convergence of the process.

Let us become more precisely. The boundary of PD(s) can be written as

$$\text{bd}(\text{PD}(s)) = [\text{bd}(\text{P}(s)) \cap \text{D}(s)] \cup [\text{P}(s) \cap \text{bd}(\text{D}(s))].$$

When the process reaches a point q in $\text{P}(s) \cap \text{bd}(\text{D}(s))$, we show that under some nondegeneracy condition either $\bar{z}(q) = \underline{0}$ or $q \in \text{D}(\bar{s})$ for some unique $\bar{s} \in \mathcal{T}$, and that q is an end point of a path in PD($\bar{s}$). Similarly, if $q \in \text{bd}(\text{P}(s))$, q will lie under some transversality condition in a set P($\hat{s}$), $\hat{s} \in \mathcal{T}$, with dimension equal to dim(P(s))-1 so that q is also an end point of a path in PD($\hat{s}$). Note that a point in bd(PD(s)) cannot be in bd(S). This because $q \in \text{PD}(s) \cap \text{bd}(S)$ implies $\bar{z}_{hk}(q) \leqq 0$ and $q_{hk} = 0$ for some $(h,k) \in I$ which contradicts condition (4.1.1.iii).

Suppose that the process in PD(s) reaches a vector $\bar{q}$ with $\text{sgn}(\bar{z}(\bar{q})) = \bar{s}$, $\bar{s} \neq \underline{0}$, in bd(D(s)). It is easy to verify that assuming nondegeneracy the following four possibilities exhaust the cases in which $\bar{s} \in \mathcal{T}$ and $\dim(\text{D}(\bar{s})) = \dim(\text{D}(s))-1$.

N1. There is an element $(0,p) \in I(0)$ such that $\bar{s}_{0p} = 0$ while $s_{0p} \neq 0$. Besides $\bar{s}_{hk} = s_{hk}$ for all $(h,k) \neq (0,p)$ with at least one element $(0,\ell) \in I(0)$ for which $s_{0\ell} \neq 0$.

N2. There is an $h \in \{1,2\}$ and just one element $(h,p) \in I(h)$ such that $\bar{s}_{hp} = 0$, $s_{hp} \neq 0$, while $\bar{s}_{j\ell} = s_{j\ell}$ for all $(j,\ell) \neq (h,p)$. Besides there is a pair $\{(h,t),(h,r)\} \subset I(h)$ such that $s_{ht} = +1$ and $s_{hr} = -1$.

N3. There are an $h \in \{0,1,2\}$ and two elements (h,t), $(h,r) \in I(h)$ such that $\bar{s}_{ht} = \bar{s}_{hr} = 0$, $s_{ht} \cdot s_{hr} = -1$ while $\bar{s}_{h\ell} = s_{h\ell} = 0$ for all $\ell \neq t,r$, and $\bar{s}_{jk} = s_{jk}$ for all other $(j,k) \in I$. In case $h = 0$, $s_{j0} = 0$ for $j \in \{1,2\}$.

N4. There are an $h \in \{1,2\}$ with $s_{h0} \neq 0$ and an element $(0,p) \in I(0)$ with $s_{0p} = s_{h0}$ while $\bar{s}_{0p} = \bar{s}_{h0} = 0$ and $\bar{s}_{0\ell} = s_{0\ell} = 0$ for all $\ell \neq p$, $\bar{s}_{j0} = s_{j0} = 0$ for $j \neq 0,h$. In case $s_{hr} \neq 0$ for exactly one element $(h,r) \neq (h,0)$ then $\bar{s}_{hr} = 0$. Furthermore, $\bar{s}_{jk} = s_{jk}$ for all other elements $(j,k) \in I$.

The next case occurs when $\bar{s} \notin \mathcal{T}$ and $\bar{z}(\bar{q}) \neq \underline{0}$.

N5. There is an element $(0,p) \in I(0)$ such that $s_{0p} \neq 0$, $\bar{s}_{0p} = 0$ and $s_{0k} = 0$ for all $k \neq p$, while $s_{10}.s_{20} = -1$. For all elements $(h,k) \in I$ with $(h,k) \neq (0,p)$, $\bar{s}_{hk} = s_{hk}$.

In case N5, $\bar{s}$ is not an element of $\mathcal{T}$, due to condition T4. Now we are able to clarify the reason for that extra condition. Suppose we define $P(\bar{s})$ according to P1-P3. But then we have that $\dim(P(\bar{s})) = \dim(P(s))$, with $\bar{s}$ and s as in N5. Hence $PD(\bar{s})$ would be a 0-manifold. However, $\bar{q}$ also lies in a set $P(\hat{s})$, with $\hat{s}_{0\ell} = -s_{0p}$ for some $\ell \neq p$, $\hat{s}_{0p} = 0$, and $\hat{s}_{hk} = s_{hk}$ for all other $(h,k) \in I$. Because $\bar{q}$ also lies in $bd(D(\hat{s}))$, the process continues from $\bar{q}$ via a path in $PD(\hat{s})$.

Next we consider the case when the vector $\bar{q}$ in $bd(PD(s))$ lies in $bd(P(s))$. We already argued that $\bar{q}$ must lie in $P(\hat{s})$, for some $\hat{s}$ in $\mathcal{T}$, with $\dim(P(\hat{s})) = \dim(P(s))-1$. This is a kind of a transversality condition which has to be fulfilled along the path. This means that it may not occur that a vector in a lower dimensional face of $bd(P(s))$ is generated. The cases that might occur are the reversals of those described in N1-N4.

4.4. Description of the price adjustments

In this section we describe how along the path the prices are adapted in order to reach an equilibrium situation. Broadly speaking, equilibrium on the domestic markets is achieved by adjustments of the domestic prices whereas equilibrium on the common market is reached via adjustments of both the international prices and the exchange rates. Adaptations of the exchange rates also restore equilibrium on the balances of payments. More precisely, an excess demand (supply) of a good induces an increase (decrease) of its price relative to the starting price. This tends to offset the imbalance. Similarly, a surplus (deficit) on the balance of payments leads to an appreciation (depreciation) of the currency in the related country. Such an appreciation (depreciation) makes for that country the common goods less (more) costly, which has also an impact on the common market. We remark that when we speak in the sequel about relative prices we mean relative to their initial values. The references P1, N1 and so on refer to Section 4.3.

At the starting vector v we assume that no balance or market is in equilibrium, i.e. all elements of $\bar{z}(v)$ differ from zero. From v, the prices of goods in excess supply are decreased while prices of goods in excess demand are increased. Besides, if a country has a surplus (deficit) on its balance of payments then its currency appreciates (depreciates) (see P1). In case all common goods reveal an excess supply (demand), their prices remain unaffected (P2 respectively P1). In that case the imbalances are attacked by appreciation (depreciation) of the national currencies.

In general, along the path prices are adjusted such that the relative price of a domestic good in excess demand is maximal, whereas the relative price of a domestic good in excess supply is minimal. Besides, the minimal relative prices are equal among the countries. Relative prices of goods in equilibrium vary between these bounds (P1). All of this also holds for the common goods prices except when no common good is in excess demand while at least one common good is not in equilibrium (see P2). Although also in that case the relative prices of common goods in excess supply are equal, they might then be larger than the relative prices of domestic goods in excess supply. In the special case that all common goods are in excess supply, their prices equal the starting prices.

If along the path a good or balance becomes in equilibrium then it is kept in equilibrium by allowing the corresponding relative price (exchange rate) to vary between the minimum and the maximum. In terms of Section 4.3 these are the cases N1 and N2 in which the boundary of a dual set has been reached. However, if the relative price (exchange rate) of a good (balance) in equilibrium becomes equal to the relative prices of the goods in excess demand or supply then the good (balance) is no longer kept in equilibrium. More precisely, its relative price (rate) is kept equal to the relative prices of the goods in excess demand or supply and the good (balance) may become in excess demand (deficit) or excess supply (surplus). This case happens when the boundary of a primal set is reached.

Till sofar we described the basic behaviour of the price adjustment process. In the sequel we treat some special cases.

Concerning country h, $h \in \{1,2\}$, it can happen that along the path two components of $\bar{z}_h$ simultaneously become equal to zero at some vector q. Under the nondegeneracy condition this can only occur when all other components of $\bar{z}_h(q)$ are zero, and hence country h becomes in equilibrium (N3). Also the reverse can occur. If a relative price (rate) in a country in equilibrium becomes equal to the relative prices of the goods in excess supply then the corresponding good (balance) is allowed to become into excess supply (surplus), while simultaneously the good (balance) with the highest relative price (rate) is allowed to become into excess demand (deficit). Similar cases can occur on the common markets when both balances are in equilibrium. But on the common markets we can also have the situation in which some goods are in equilibrium while all other common goods are either in excess supply (P2) or in excess demand (P1). In the latter case the adaptations are standard. In the former case the relative prices of the common goods in excess supply are less than or equal to the relative prices of the common goods in equilibrium, but they may be higher than the relative prices of the domestic goods in excess supply. If the relative prices of the common goods in excess supply become equal to the relative prices of the non-common goods in excess supply, then the former relative prices are kept equal to the latter while the common good in equilibrium having the highest relative price is allowed to become into excess demand (change from P2 to P1). Of course the opposite situation occurs when the last common good in excess demand becomes in equilibrium (N1).

Next, we have to consider the cases in which, due to the connections between the common markets and the balances of the countries (4.1.5b), simultaneously changes occur at the common markets and the balances. First, it can happen that the last common good and the last balance not being in equilibrium both become in equilibrium (N4). Then the corresponding price and rate are adjusted such as to keep them in equilibrium. Of course, also the opposite case is possible. In that case a rate becomes equal to the relative prices of the domestic goods in excess supply (demand). Then the rate is kept equal to this relative price and the balance may reveal a surplus (deficit). Simultaneously, the common good with the lowest (highest) relative price may become into excess supply (demand). Finally, it can occur that the last common good not being in equilibrium, becomes in equilibrium while both balances are not in equilibrium, i.e. there is a balance revealing a surplus while the other is in deficit (N5). When that common good was in excess supply (demand), the process continues by allowing the common good with the highest (lowest) relative price to become into excess demand (supply).

This completes the description of the price adjustments made by our process. The most remarkable feature of the process in relation to other price adjustment processes is that it focusses on relative prices, i.e. prices relative to the starting price system. In this way the starting vector plays a very important role. At any point along the path the process keeps track of its position with respect to that starting vector. If necessary, this leads to disturbances of partial equilibria. The advantage resulting from this is that the process converges to an equilibrium under rather general conditions. Furthermore, the process describes economically appealing price adjustments for an international trade model including adjustments of the exchange rates.

4.5. Numerical illustration

We consider a simple international trade model with two countries, 1 and 2. The consumers in each country are represented by one single consumer. Furthermore, each country has one domestic good, indexed (1,1) and (2,1) respectively, and there are two commonly traded goods, indexed (0,0) and (0,1). The representative consumer in country 1 maximizes a Cobb-Douglas utility function $u^1(x_{00},x_{01},x_{11}) = x_{00}^{\alpha_1} x_{01}^{\alpha_2} x_{11}^{\alpha_3}$, with $\alpha_i \geq 0$,

$i \in \{1,2,3\}$, and $\alpha_1+\alpha_2+\alpha_3 = 1$. Here x_{00}, x_{01} and x_{11} denote the quantities consumed. Similarly, we have a utility function for the consumer in country 2, $u^2(x_{00},x_{01},x_{21}) = x_{00}^{\beta_1} x_{01}^{\beta_2} x_{21}^{\beta_3}$, with $\beta_i \geq 0$, $i \in \{1,2,3\}$, and $\beta_1+\beta_2+\beta_3 = 1$. The budget constraints are induced by the initial endowments of the countries. For country 1 we have endowments $w^1 = (w^1_{00}, w^1_{01}, w_{11})^T$, and for country 2, $w^2 = (w^2_{00}, w^2_{01}, w_{21})^T$. In our numerical example we take $\alpha_i = \beta_i = 1/3$, $i \in \{1,2,3\}$. Note that the excess demand is not defined for zero prices. In that case the demands become infinite. This does not matter because our process never reaches zero prices (see Section 4.3).

On this model we perform the following experiment. First, we consider the way in which our process restores the equilibrium after the occurrence of a shock. We start the process in the old equilibrium price vector and consider the adaptations leading to the new equilibrium. After that we observe the consequences of the reversed shock. We then compare the two paths.

In the starting situation of the economy we take $w^1 = (100,60,80)^T$ and $w^2 = (0,40,20)^T$. The corresponding equilibrium vector $q^* = ((q^*_{00},q^*_{01})^T,(q^*_{10},q^*_{11})^T,(q^*_{20},q^*_{21})^T)$ equals $((1/2,1/2)^T,(2/3,1/3)^T,(2/3,1/3)^T)$. Now, a change in the initial endowments triggers off a shock in the economy. More concrete, w^1 becomes $(0,60,180)^T$ whereas w^2 becomes $(50,40,50)^T$. Straightforward computation yields that the equilibrium vector $\bar{q}$ corresponding to this new situation is equal to $\bar{q} = ((2/3,1/3)^T, (18/19,1/19)^T,(15/22,7/22)^T)$. We now consider the adjustment process from the old equilibrium prices q^* to $\bar{q}$. Let $\bar{z}$ be the excess demand function in the new situation. Then $\bar{z}(q^*)$ equals $(\bar{z}_{00}(q^*),\bar{z}_{01}(q^*),\bar{z}_{10}(q^*),\bar{z}_{11}(q^*), \bar{z}_{20}(q^*),\bar{z}_{21}(q^*))^T = (76.54,26.54,49.92,-99.84,1.62,-3.24)^T$. So, at the start, both common goods are in excess demand, the balances are in deficit and the domestic goods are in excess supply. Thus, the process leaves q^* by devaluating the domestic currencies and decreasing equally the domestic goods prices while the international prices of the common goods are kept fixed. The adjustment process continues in this manner till the vector $((0.5,0.5)^T,(0.689,0.311)^T,(0.689,0.311)^T)$ is reached at which country 2 becomes in equilibrium. The latter means that both the domestic market of country 2 and its balance become in equilibrium. From that vector on, country 2 is kept in equilibrium whereas the other adaptations are continued as before. Then, at the vector $((0.5,0.5)^T,(0.774,0.226)^T,(0.689,0.311)^T)$, the second common good market becomes in equilibrium. The market for the first common good still reveals a surplus. Now, the process proceeds while also keeping the second common good in equilibrium and increasing the price of the first common good above the price of the second one till the new equilibrium vector $\bar{q}$ is reached. At $\bar{q}$ also country 1 becomes in equilibrium which also induces equilibrium on the first common market.

Next we consider the reverse case. We start from the situation with $w^1 = (0,60,180)^T$, $w^2 = (50,40,50)^T$, and corresponding equilibrium $\bar{q} = ((2/3,1/3)^T,(18/19,1/19)^T,(15/22,7/22)^T)$ and assume a shock on this economy changing the endowments back to $w^1 = (100,60,80)^T$ and $w^2 = (0,40,20)^T$. How does our process adjust $\bar{q}$ to the new equilibrium $q^* = ((1/2,1/2)^T,(2/3,1/3)^T,(2/3,1/3)^T)$? The excess demand vector at $\bar{q}$ after the shock equals $\bar{z}(\bar{q}) = (-43.17,13.66,-25.99,467.76,1.76,-3.78)^T$. Thus, the shock results for country 1 in an excess demand situation on its domestic market whereas its balance turns into surplus. In country 2 the opposite occurs, i.e. its balance runs into deficit whereas the domestic market becomes in excess supply. Furthermore, the market for the first common good is in excess supply while the other common market is in excess demand. The process leaves $\bar{q}$ by proportionally decreasing the first common good price and the domestic price in country 2 while relatively revaluating the currency of country 1 with the same factor. These adaptations are continued till country 2 becomes in equilibrium at the price vector $((0.599,0.401)^T,(0.852,0.148)^T,(0.714,0.286)^T)$. From that vector on, country 2 is kept in equilibrium by adjusting the ratio between the price of the domestic good and its exchange rate whereas the other adjustments are continued as before. Next, the process reaches the vector $((0.585,0.415)^T,(0.832,0.168)^T,(0.705,0.295)^T)$ at which the market for the second common good reveals an equilibrium. By keeping this market and country 2 in equilibrium, increasing the first common good price relatively above the exchange rate in country 1, and simultaneously continuing the previous adaptations the new equilibrium q^* is reached. At the latter vector also country 1 becomes in equilibrium and henceforth the market for the first common good.

It is interesting to observe from the experiment above that the path traced by the process from the second to the first equilibrium is not the reverse of the path followed from the first to the second equilibrium situation. This appears to be a general feature of the process. It is mainly due to the fact that the price decreases are, relatively to the starting vector, equal among the markets. The differences occur therefore because of the different starting vectors. Another fundamental reason is that the process leaves the starting vector along a ray whereas the equilibrium vector is reached via a curve.

CHAPTER 5

An adjustment process for an exchange economy with linear production technologies

In this chapter we propose a process that reaches an equilibrium in an exchange economy with linear production via adaptations of prices and activity levels. The process we consider here is a generalization of the sign process for a pure exchange economy (see van der Laan and Talman [1987a] and Section 3.1 of Chapter 3). It is a generalization in the sense that for the special case of an exchange economy without production both processes are the same.

We remark that we discuss a process, not an algorithm. A process is a theoretical device whereas an algorithm is a tool for finding an equilibrium. The standard algorithms for finding an equilibrium in an exchange economy with linear production are those of Mathiesen [1985a] and Eaves [1987]. They solve the problem as a sequence of linear complementarity problems. Kremers [1991] gives an algorithm which solves the problem by solving a sequence of linear stationary point problems. Finally, in Kremers [1991] a simplicial algorithm is given. The path of points generated by that algorithm can follow the path of the process presented here arbitrarily close. Another appropriate simplicial algorithm can be found in Talman and Yamamoto [1989].

This chapter consists of four sections. In Section 5.1 we present a model of an exchange economy with linear production technologies and introduce the process. In Section 5.2 we give the mathematical description of the path followed by the process and prove its existence and convergence. The economic interpretation of the path is given in Section 5.3. Finally, in Section 5.4 we consider two examples. This chapter is based on van den Elzen, van der Laan, and Talman [1990].

5.1. The model

We consider a standard model of an exchange economy with linear production technologies. There are a finite number of consumers, m production activities, also called firms, and n+1 commodities. The firms are indexed by i, $i \in \{1,\ldots,m\}$, and the commodities or goods by j, $j \in \{1,\ldots,n+1\}$. Consumers are assumed to be endowed with some of the goods, such as labour and capital. More precisely, the nonnegative vector w in $\mathbb{R}^{n+1}$ denotes the (aggregated) initial endowments of the consumers, with w_j the amount of commodity j, $j \in \{1,\ldots,n+1\}$. At a price vector p in $\mathbb{R}^{n+1}_+\backslash\{\underline{0}\}$, the (n+1)-vector d(p) denotes the aggregate demand of the consumers for the commodities. We assume that demand is homogeneous of degree zero in the prices and that it satisfies Walras' law, i.e. $d(\lambda p) = d(p)$ for all $\lambda > 0$ and $p \cdot d(p) = p \cdot w$, respectively, for every price vector p. We also assume that the function d is smooth on $\mathbb{R}^{n+1}_{++}$ and continuous on $\mathrm{bd}(\mathbb{R}^{n+1}_+)\backslash\{\underline{0}\}$.

An activity of a firm is represented by an (n+1)-vector whose negative components correspond to the inputs and whose positive components to outputs. More precisely, the (n+1)-vector a^i denotes the activity vector of firm i, $i \in I_m$, with $a^i_j \geqq 0$ the (net) amount of output of commodity j and $-a^i_j \geqq 0$ the amount of input of commodity j, $j \in I_{n+1}$, when the activity level is equal to one. A vector y in $\mathbb{R}^m_+$ denotes an activity level vector with y_i the activity level of firm i, $i \in I_m$. So, the (n+1)-vector Ay, with A the $((n+1)\times m)$-matrix with i-th column a^i for $i \in I_m$, is the aggregate net input-output vector for activity level vector y. We assume that there can be no production without input, i.e. $Ay \geqq \underline{0}$ and $y \geqq \underline{0}$ imply $y = \underline{0}$.

The m-vector $A^{\mathsf{T}}p$ is the unit level profit vector at price vector p, with $(A^{\mathsf{T}}p)_i$ the profit of firm i when it operates at unit level. We call a price vector p^* and an activity level vector y^* an equilibrium if for each commodity the demand is at most equal to endowment plus net production and if no activity makes profit.

<u>Definition 5.1.1</u>. A pair $(p^*,y^*) \in (\mathbb{R}^{n+1}_+\backslash\{\underline{0}\}) \times \mathbb{R}^m_+$ is an equilibrium if

i) $\quad d(p^*) - Ay^* \leqq w$

ii) $A^Tp^* \leqq \underline{0}$.

From the definition and Walras' law we can derive some properties holding at an equilibrium. Multiplying i) with p^* and ii) with y^* yields $-p^*\cdot Ay^* \leqq 0$ and $p^*\cdot Ay^* \leqq 0$, respectively. Thus, $p^*\cdot Ay^* = 0$ and therefore with Walras' law also $p^*\cdot(d(p^*) - Ay^* - w) = 0$. Because $A^Tp^* \leqq \underline{0}$ and $y^* \geqq \underline{0}$, $p^*\cdot Ay^* = 0$ means that in equilibrium a firm can only operate at a positive production level if it makes zero profit, i.e. if $y_i^* > 0$ then $(A^Tp^*)_i = 0$. Similarly, because $d(p^*) - Ay^* - w \leqq \underline{0}$ holds, $p^*\cdot(d(p^*) - Ay^* - w) = 0$ means that in equilibrium the consumers' demand for a commodity can only be less than its endowments plus net production if the price of that commodity is equal to zero.

Due to the homogeneity of degree zero of the demand function d we have that if (p^*,y^*) is an equilibrium, then also $(\lambda p^*,y^*)$ is an equilibrium for any $\lambda > 0$. This allows us to normalize the price vectors to the n-dimensional unit simplex S^n. Because d is a smooth function on the interior of S^n and continuous on $bd(S^n)$, it is well-known from a fixed-point argument that an equilibrium always exists in $S^n\times\mathbb{R}^m_+$ (see e.g. Scarf [1973]). Since at an equilibrium $(p^*,y^*) \in S^n\times\mathbb{R}^m_+$ it holds that $A^Tp^* \leqq \underline{0}$, p^* lies in S^n_A, where $S^n_A := \{p \in S^n | A^Tp \leqq \underline{0}\}$. To find an equilibrium vector pair (p^*,y^*) in $S^n_A\times\mathbb{R}^m_+$, we propose to follow a piecewise smooth path, denoted P, in $S^n_A\times\mathbb{R}^m_+$. Let p^0 be an arbitrarily chosen point in the interior of S^n_A, i.e. $p^0_j > 0$ for all j and $p^0\cdot a^i < 0$ for all i. According to Farkas' lemma the relative interior of S^n_A is nonempty, since $Ay \geqq \underline{0}$ and $y \geqq \underline{0}$ imply $y = \underline{0}$. The path connects the pair $(p^0,\underline{0})$ and an equilibrium pair (p^*,y^*) and can be interpreted as a path generated by an adjustment process in which prices and activity levels simultaneously adjust. All points (p,y) along the path P in $S^n_A\times\mathbb{R}^m_+$ satisfy for $j \in I_{n+1}$

$$p_j/p_j^0 = \min_h p_h/p_h^0 \qquad \text{if } d_j(p) - (Ay)_j < w_j$$

$$\min_h p_h/p_h^0 \leqq p_j/p_j^0 \leqq \max_h p_h/p_h^0 \qquad \text{if } d_j(p) - (Ay)_j = w_j$$

$$p_j/p_j^0 = \max_h p_h/p_h^0 \qquad \text{if } d_j(p) - (Ay)_j > w_j,$$

and for $i \in I_m$ (5.1.1)

$$y_i = 0 \qquad \text{if } p \cdot a^i < 0$$

$$y_i \geqq 0 \qquad \text{if } p \cdot a^i = 0.$$

5.2. Existence of the path

An arbitrary price vector p^0 in $\text{int}(S^n_A)$ can be determined by solving the Linear Programming Problem (LPP)

$$c = \min(p_1 + \ldots + p_{n+1})$$

$$\text{such that } A^T p \leqq -e^m \text{ and } p \geqq e^{n+1}.$$

Clearly, this LPP has a solution $\bar{p}$. After dividing $\bar{p}$ by c, the sum of its components, a price vector p^0 in the interior of S^n_A is obtained.

Given p^0, the set of points $(p,y) \in S^n_A \times \mathbb{R}^m_+$ satisfying (5.1.1) is denoted by B, i.e. B is the set of points (p,y) in $S^n_A \times \mathbb{R}^m_+$ such that

i) for $j \in I_{n+1}$,

$$p_j/p^0_j = \min_h p_h/p^0_h \qquad \text{if } d_j(p) - (Ay)_j < w_j$$

and

$$p_j/p^0_j = \max_h p_h/p^0_h \qquad \text{if } d_j(p) - (Ay)_j > w_j,$$

ii) $y_i = 0$ if $p \cdot a^i < 0$, $i \in I_m$.

Clearly, the point $(p^0,\underline{0})$ satisfies i) and ii) with both the minimum and the maximum equal to one. Also, all equilibria lie in the set B. We now show that under standard nondegeneracy and transversality conditions the set B consists of piecewise smooth paths and loops. One of these paths is the path P, connecting $(p^0,\underline{0})$ and an equilibrium (p^*,y^*).

Observe that at all $(p,y) \in B$ condition ii) of Definition 5.1.1 is fulfilled. If also $d(p) - Ay - w \leqq \underline{0}$ then (p,y) is an equilibrium. Furthermore, for all $(p,y) \in B$ we have that that $p \cdot Ay = 0$ because $y_i = 0$ if $p \cdot a^i < 0$, $i \in \{1,\ldots,m\}$. Hence according to Walras' law we also have that

$p\cdot(d(p) - Ay - w) = 0$. Suppose now that there exists an index j for which $d_j(p) - (Ay)_j - w_j > 0$. Since $(p,y) \in B$ then $p_j = p_j^0.\max_h p_h/p_h^0 > 0$. Therefore, $d(p) - Ay - w$ must also contain at least one negative component with corresponding price positive. In the sequel of this chapter s denotes a sign vector in $\mathbb{R}^{n+1}$. If s contains at least one +1 and one -1 we say that s is a <u>feasible</u> sign vector.

Let U be a subset of I_m. We want to split up B into subsets related to a feasible sign vector s and a subset U. For that we first define $A(s,U)$ by

$$A(s,U) = \{(p,y) \in S_A^n \times \mathbb{R}_+^m \mid a^i\cdot p = 0 \text{ for } i \in U,\ y_i = 0 \text{ for } i \notin U,$$

$$p_j/p_j^0 = \min_h p_h/p_h^0 \quad \text{when } s_j = -1$$

$$p_j/p_j^0 = \max_h p_h/p_h^0 \quad \text{when } s_j = +1\}.$$

Note that there are $n-1-|I^0(s)|+|U|$ constraints on $p \in S_A^n$ and $m-|U|$ constraints on y. Thus, if $|I^0(s)| = |U|$, there are n-1 constraints on p and because $\dim(S_A^n)$ equals n, one degree of freedom is left. Indeed, the p-components in a nonempty $A(s,U)$ then form a line segment and so a set of dimension 1. Clearly, if $|U| > |I^0(s)|+1$ there are more than n constraints on p and no price vector in S_A^n satisfies all conditions, i.e. $A(s,U) = \emptyset$. If $|U| \leqq |I^0(s)|+1$, a nonempty set $A(s,U)$ is well-defined and its dimension is equal to $|I^0(s)|+1$.

We now consider the set of points (p,y) satisfying $(p,y) \in A(s,U)$ for $s = \text{sgn}(d(p) - Ay - w)$. We denote the closure of this set by $B(s,U)$. Clearly, every point (p,y) satisfying (5.1.1) lies in some set $B(s,U)$. More precisely, B is the union of $B(s,U)$ over all feasible sign vectors s and subsets U of I_m with $|U| \leqq |I^0(s)|+1$. In particular the point $(p^0,\underline{0})$ lies in $A(s^0,\emptyset)$, where $s^0 = \text{sgn}(d(p^0) - w)$. Without loss of generality we assume that the excess demand vector $d(p^0) - w$ does not contain zero components so that the dimension of $A(s^0,\emptyset)$ equals 1. Observe that in the definition of each $B(s,U)$ one degree of freedom is left because $\dim(A(s,U)) = |I^0(s)|+1$ whereas the condition that $\text{sgn}(d(p) - Ay - w) = s$ imposes $|I^0(s)|$ conditions on (p,y). Thus, assuming standard nondegeneracy and transversality conditions, a nonempty $B(s,U)$ forms a collection of disjoint smooth paths and loops. An end point (p,y) of a path in $B(s,U)$ is characterized by one of the following cases:

0) $(p,y) = (p^0,\underline{0})$;

i) $\min_h p_h/p_h^0 = 0$;

ii) $p_j/p_j^0 = \min_h p_h/p_h^0$ for some $j \in I^0(s)$;

iii) $p_j/p_j^0 = \max_h p_h/p_h^0$ for some $j \in I^0(s)$; (5.2.1)

iv) $a^i \cdot p = 0$ for some $i \notin U$;

v) $y_i = 0$ for some $i \in U$;

vi) $d_j(p)-(Ay)_j = w_j$ for some $j \notin I^0(s)$.

We argue that an end point of a path in $B(s,U)$ is either $(p^0,\underline{0})$, or an equilibrium, or an end point of a unique path in some other set $B(\bar{s},\bar{U})$, so that the paths in different sets $B(s,U)$ can be linked to form piecewise smooth paths and loops.

Suppose that case 0) occurs. Since $d(p^0) - w$ is assumed not to contain zeros and $(A^{\top}p^0)_i < 0$ for all i, s must be equal to s^0 and $U = \emptyset$. Thus, $(p^0,\underline{0})$ is only an end point of one path in $B(s^0,\emptyset)$.

In case i) we must have that $p_j = 0$ for all j for which $s_j = -1$. Hence, $d_j(p)-(Ay)_j \leqq w_j$ for all indices j for which $p_j = 0$, whereas $d_j(p)-(Ay)_j \geqq w_j$ whenever $p_j > 0$. Since, $p\cdot(d(p) - Ay - w) = 0$ for all $(p,y) \in B$, this implies $d_j(p)-(Ay)_j = w_j$ if $p_j > 0$. Therefore (p,y) is an equilibrium.

Next suppose that case ii) or case iii) holds. Then (p,y) must be also an end point of a path in $B(\bar{s},U)$ where $\bar{s}_h = s_h$ for all $h \neq j$ and $\bar{s}_j = -1$ in case ii), $\bar{s}_j = +1$ in case iii). Clearly, this path is uniquely determined.

In case iv) the point (p,y) is also an end point of a path in $B(s,U \cup \{i\})$ and in case v) of a path in $B(s,U\backslash\{i\})$. In both cases these paths are uniquely determined.

Finally, we consider case vi). If $s_j = +1$ (-1) and s has no other positive (negative) components, then according to the fact that $p \cdot (d(p) - Ay - w) = 0$, (p,y) must be an equilibrium. Otherwise (p,y) is also an end point of a path in $B(s',U)$ where $s'_j = 0$ and $s'_h = s_h$ for all $h \neq j$. Again, this path is uniquely determined.

Consequently, for different s and U, the paths in the sets $B(s,U)$ can be linked to form disjoint piecewise smooth paths and loops. Exactly one path P has $(p^0,\underline{0})$ as an end point. All other end points of these paths are equilibria. To prove that the path P has another end point, which must be then an equilibrium, we show in the next lemma that the set B is bounded.

Lemma 5.2.1. The set B is bounded.

Proof. Suppose that the set B is unbounded. Then without loss of generality there is some feasible sign vector s and set U such that $B(s,U)$ contains a sequence $\{(p^k,y^k)\}_{k=1}^{\infty}$, with some of the components of (p^k,y^k) going to infinity. Since S_A^n is compact the sequence $\{p^k\}_{k=1}^{\infty}$ is bounded and has a cluster point, $\hat{p}$. Hence, some of the components of y^k must go to infinity. Because $(p^k,y^k) \in B(s,U)$ for each k, there exist $\mu_h^k \geqq 0$, $h \notin I^0(s)$, such that

$$d(p^k) - \sum_{i \in U} a^i y_i^k - \sum_{s_h \neq 0} \mu_h^k s_h e(h) = w.$$

Since p^k converges to $\hat{p}$ and d is continuous, the latter system can only have a solution for all k if the homogeneous system of linear equations

$$\sum_{i \in U} a^i y_i + \sum_{s_h \neq 0} \mu_h s_h e(h) = \underline{0} \tag{5.2.2}$$

has a nonzero solution $\bar{y}_i \geqq 0$ for $i \in U$ and $\bar{\mu}_h \geqq 0$ for $h \notin I^0(s)$. Since $(p^k,y^k) \in B(s,U)$ for all k, there exist a number b, $0 < b < 1$, and a vector $q \in \mathbb{R}_+^{n+1}$ with $q_j > 0$ if $s_j = +1$, $q_j \geqq 0$ if $s_j = 0$, and $q_j = 0$ if $s_j = -1$ such that $\hat{p} = bp^0 + q$. Clearly, $q \cdot a^i > 0$ for $i \in U$ since $\hat{p} \cdot a^i = 0$ and $p^0 \cdot a^i < 0$ for $i \in U$. Premultiplying (5.2.2) with q^T yields

$$\sum_{i \in U} (q \cdot a^i) y_i + \sum_{s_h = +1} \mu_h q_h = 0.$$

Since $q \cdot a^i > 0$ for $i \in U$ and $q_h > 0$ for $s_h = +1$, this can only hold when all the y_i's and all μ_h's for which $s_h = +1$ are equal to zero. But then according to (5.2.2) all other μ_h's must also be equal to zero. Hence, system (5.2.2) has no nonzero nonnegative solution, which completes the proof.

□

Lemma 5.2.1 implies that the path P is bounded and therefore has another end point which must be an equilibrium.

5.3. The adjustment process

In this section we provide an economic interpretation of the adjustments of prices and activity levels along the path P as defined in Section 5.2. The references made in the text are to the cases listed in (5.2.1). In the sequel we mean p_j/p_j^0, $j \in I_{n+1}$, when we speak about the price ratio of commodity j at a price vector p. Further, we use the following notation. By z(p) we denote the consumers' excess demand at price vector p, i.e. $z(p) := d(p) - w$, whereas $\tilde{z}(p,y)$ denotes the (total) excess demand at price vector p and activity level vector y, i.e. $\tilde{z}(p,y) := z(p) - Ay$.

The process starts in (p^0,y^0) with $y^0 = \underline{0}$ and a price vector p^0 such that all prices are positive and all activities make losses. Moreover, at p^0, the excess demand $\tilde{z}(p^0,\underline{0})$ is equal to the consumers' excess demand $z(p^0)$ and is assumed not to contain zeros. Now, the process leaves $(p^0,\underline{0})$ by increasing proportionally the prices of the commodities in excess demand ($z_j(p^0) > 0$) and decreasing proportionally the prices of the commodities in excess supply ($z_j(p^0) < 0$). The process continues in this way until a price vector p is reached at which either a price becomes zero (case i)) or one of the goods becomes in equilibrium, i.e. $z_j(p) = 0$ for some j (case vi)), or one of the activities makes zero profit, i.e. $p \cdot a^i = 0$ for some i (case iv)). In case i) an equilibrium has been reached because in this case all the prices of the goods in excess supply have become zero simultaneously. Walras' law then implies that there cannot be a good in excess demand anymore. In case vi) the process continues by keeping the price ratios of the goods in excess demand (excess supply) maximal (minimal) while the good j is kept in equilibrium by varying its price ratio. Finally, in case iv) the activity level y_i of activity i is increased from zero. This activity level is increased until the excess demand or supply of some commodity becomes zero. Lemma 5.2.1, saying that the set of points (p,y) in B is bounded, guarantees that this will occur as can also be seen as follows. Let $s^0 = \text{sgn}(z(p^0))$ then also at p we must have $\text{sgn}(z(p)) = s^0$. Since $p \cdot a^i = 0$, $p^0 \cdot a^i < 0$, $p_j > p_j^0$ if $s_j^0 = +1$, and $p_j < p_j^0$ if $s_j^0 = -1$, there exists at least one index h such that $\text{sgn}(a_h^i) = s_h^0$. From this it immediately follows that $d_h(p) - a_h^i y_i - w_h$ must go to zero for at least one index h if y_i is increased from zero.

In general, the process generates a path of price vectors p in S_A^n and activity level vectors y in $\mathbb{R}_+^m$. At such pair (p,y) the price ratio p_j/p_j^0 of a good in excess demand ($\tilde{z}_j(p,y) > 0$) is maximal and the price ratio of a good in excess supply ($\tilde{z}_j(p,y) < 0$) is minimal. The price ratio of a good in equilibrium ($\tilde{z}_j(p,y) = 0$) lies between this minimum and maximum. An activity level can only be positive ($y_i > 0$) if the corresponding activity makes zero profit ($p \cdot a^i = 0$). Finally, an activity not operating ($y_i = 0$) makes negative or zero profit ($p \cdot a^i \leqq 0$).

As soon as a pair (p,y) is reached at which a good becomes in equilibrium, that good is kept in equilibrium and its price ratio is allowed to vary between the minimal and maximal price ratios. On the other hand, if the price ratio of a good in equilibrium becomes equal to the maximal (or minimal) price ratio, then it is kept equal to the maximal (minimal) price ratio and the good becomes in excess demand (excess supply).

When a pair (p,y) is reached at which an activity not producing makes zero profit, then the level of this activity is allowed to become positive and its profit is kept equal to zero. Also, when the process reaches a (p,y) at which the level of an activity making zero profit becomes zero, then the process continues by keeping this activity level equal to zero while the price adaptations are allowed to bring this activity into a loss situation.

The process stops as soon as each market is in equilibrium or in excess supply. Lemma 5.2.1 guarantees that the process indeed will reach an equilibrium.

In the interpretation given above, the adaptations of prices and activity levels were treated more or less separately. In fact, they adjust simultaneously. Changes in activity levels do have an impact on excess demands, whereas changes in price levels influence both all the excess demands and profits. We illustrate this with a general example. The path generated by the process is depicted in Figure 5.3.1. This figure shows the projection of the path on the price space. The reader should keep in mind that the process adjusts prices as well as activity levels and in fact operates in $S^n_A \times \mathbb{R}^m_+$ and not in S^n_A.

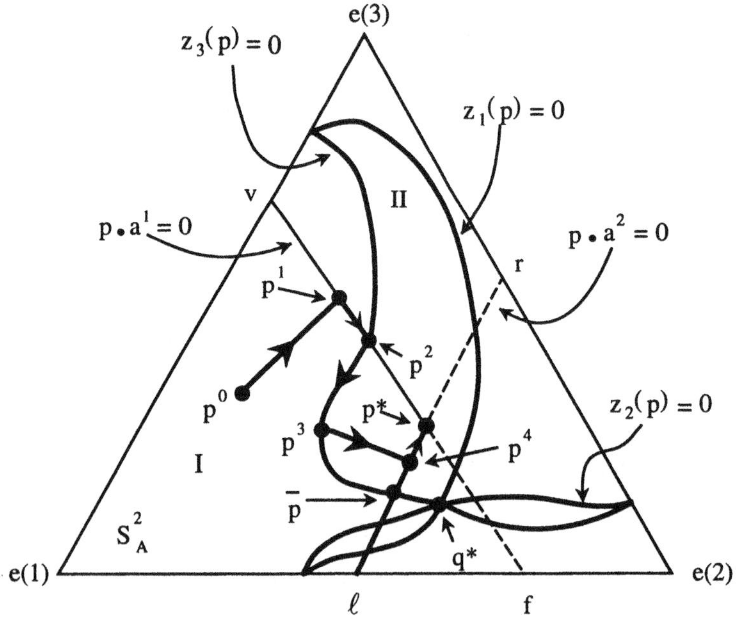

Figure 5.3.1. Exchange economy with three goods and two activities. At prices p in region I, $\mathrm{sgn}(z(p)) = (-1,+1,+1)^T$. For region II holds $\mathrm{sgn}(z(p)) = (-1,+1,-1)^T$.

We consider an economy with three goods and two activities. All relevant information of the economy is given in Figure 5.3.1. The simplex is the price set. Activity 1 uses the first commodity as input whereas the other goods are outputs. We denote this as $a^1 = (-,+,+)^\top$. Similarly, we have $a^2 = (-,+,-)^\top$. The line segments $[v,f]$ and $[\ell,r]$ denote the price vectors at which the profit of activity 1 and 2, respectively, is zero. Because at $p = e(1)$ both activities make losses the set S^2_A is the convex hull of $e(1)$, v, p^* and ℓ. Also drawn are the curves $\{p|z_j(p) = 0\}$, $j \in \{1,2,3\}$. We assume that $z_j(p)$ is positive whenever $p_j = 0$. Because of Walras' law the curves $\{p|z_j(p) = 0\}$ and $\{p|z_h(p) = 0\}$, $j \neq h$, meet each other on the edge $\{p|p_i = 0,\ i \neq j,h\}$. Moreover, the three curves intersect each other in q^* which is the equilibrium of the corresponding pure exchange economy. Note that q^* lies outside the region S^2_A, so that the economy with production activities has no equilibrium for zero activity levels.

Let us consider now what happens along the path of the process starting from $(p^0,\underline{0})$. Since p^0 lies in the interior of S^2_A both activities make losses. At $(p^0,\underline{0})$ commodity 1 is in excess supply, and the commodities 2 and 3 are in excess demand. The process leaves $(p^0,\underline{0})$ by proportionally increasing the prices of goods 2 and 3 and decreasing the price of good 1. Thus, the path of prices leaves p^0 in the direction opposite to $e(1)$. Meanwhile, the activity levels remain zero.

At $(p^1,\underline{0})$ the profit of activity 1 becomes zero and its activity level y_1 is increased. Because $\text{sgn}(\tilde{z}(p^1,\underline{0})) = \text{sgn}(z(p^1)) = (-1,+1,+1)^\top$ and $a^1 = (-,+,+)^\top$, this increase tends to offset the imbalances on all markets. The level y_1 is increased till say $\bar{y}_1$, at which one of the markets becomes in equilibrium. In case commodity 1 would become in equilibrium the other markets must then also be in equilibrium and hence $(p^1,(\bar{y}_1,\underline{0})^\top)$ is an equilibrium. This because $p\cdot\tilde{z}(p,y) = 0$ for all (p,y) on the path whereas $\tilde{z}_2(p^1,\underline{0}) > 0$ and $\tilde{z}_3(p^1,\underline{0}) > 0$. If commodity 2 would become in equilibrium, i.e. $\tilde{z}_2(p^1,(\bar{y}_1,0)^\top) = 0$, then the prices and y_1 are adjusted simultaneously such that commodity 2 is kept in equilibrium, the profit of activity 1 remains zero, and the price ratio of commodity 2 lies between the minimum price ratio (p_1/p^0_1) and the maximum price ratio (p_3/p^0_3). In the figure the path of prices would then move in the direction of v. It can be shown that then there would be an equilibrium $(\tilde{p},(\tilde{y}_1,0)^\top)$ with $\tilde{p}$ on the line segment $[p^1,v]$. Finally it may happen that at $(p^1,(\bar{y}_1,0)^T)$ commodity 3 becomes in equilibrium ($\tilde{z}_3(p^1,(\bar{y}_1,0)^T) = 0$). Then the prices and the activity level y_1 are adjusted simultaneously such as to keep commodity 3 in equilibrium whereas the profit of activity 1 remains zero and the price ratio of good 3 (p_3/p_3^0) varies between the minimum price ratio (p_1/p_1^0) and the maximum price ratio (p_2/p_2^0). In the figure the path of prices moves towards f. If between p^1 and p^2 the market of commodity 1 would become in equilibrium then also the market of commodity 2 must become in equilibrium and an equilibrium of the economy is reached. Again this is due to the fact that $p \cdot \tilde{z}(p,y) = 0$ along the path.

Suppose there is no equilibrium between p^1 and p^2. Then the process reaches p^2 where, because $z_3(p^2) = 0$, the level of activity 1 needed to keep commodity 3 in equilibrium must be equal to zero. Now, y_1 is fixed at 0, prices are allowed to bring activity 1 into a loss situation whereas good 3 is still kept in equilibrium. Thus, the process follows a path of vectors $(p,\underline{0})$ with p in $\text{int}(S_A^2)$ and $z_3(p) = 0$. In the figure this piece of the path is represented by the curve connecting p^2 and p^3.

Then, at p^3 the price ratio of commodity 3 (p_3^3/p_3^0) becomes equal to the minimum price ratio (p_1^3/p_1^0). This because p^3 lies on the line segment connecting p^0 and e(2). The process continues from $(p^3,\underline{0})$ by keeping the price ratios of commodities 1 and 3 equal to each other ($p_3/p_3^0 = p_1/p_1^0$) whereas commodity 3 is allowed to become into excess supply. In Figure 5.3.1, the path of prices enters region II in the direction of e(2).

At $(p^4,\underline{0})$ the profit of activity 2 becomes zero and its activity level y_2 is increased. This situation is similar to that at $(p^1,\underline{0})$. Because $\text{sgn}(\tilde{z}(p^4,\underline{0})) = (-1,+1,-1)^T$ whereas $a^2 = (-,+,-)^T$, the increase of y_2 diminishes both the excess demand for good 2 and the excess supplies of commodities 1 and 3. Thus, there is a level $\bar{y}_2$ at which one of the commodities becomes in equilibrium. If $\tilde{z}_2(p^4,(0,\bar{y}_2)^T) = 0$ then $(p^4,(0,\bar{y}_2)^T)$ would be an equilibrium because of Walras' law. In case $\tilde{z}_1(p^4,(0,\bar{y}_2)^T) = 0$ then prices and y_2 would be adjusted such that commodity 1 is kept in equilibrium, the profit of activity 2 remains zero whereas the price ratio p_1/p_1^0 varies between the maximum price ratio (p_2/p_2^0) and the minimum price ratio (p_3/p_3^0). The corresponding path of prices in Figure 5.3.1 would go in the direction of ℓ. We argue that in that case there would be an equilibrium on the open segment $(p^4,\bar{p})$. Assume the contrary. Then the process would reach $\bar{p}$ with corresponding $\hat{y}_2$ such that $\tilde{z}_1(\bar{p},(0,\hat{y}_2)^T) = 0$. Note that $\hat{y}_2 > 0$ because $z_1(\bar{p}) < 0$. However, because $z_3(\bar{p}) = 0$, we then have that $\tilde{z}_3(\bar{p},(0,\hat{y}_2)^T) > 0$. Thus, the sign of $\tilde{z}_3$ changes from -1 at $(p^4,(0,\bar{y}_2)^T)$ into +1 at $(\bar{p},(0,\hat{y}_2)^T)$. But then there must be a point $(p,(0,y_2)^T)$ with $p \in (p^4,\bar{p})$, at which $\tilde{z}_3(p,(0,y_2)^T) = \tilde{z}_1(p,(0,y_2)^T) = 0$, and hence $\tilde{z}_2(p,(0,y_2)^T) = 0$, i.e. $(p,(0,y_2)^T)$ would be an equilibrium. Finally, we consider the case when at $(p^4,(0,\bar{y}_2)^T)$ commodity 3 becomes in equilibrium. Then the process keeps commodity 3 in equilibrium and fixes the profit of activity 2 at zero while varying the price ratio of commodity 3 between the minimum price ratio (p_1/p_1^0) and the maximum price ratio (p_2/p_2^0). In Figure 5.3.1 the path projected on the price space moves from p^4 towards r. If the process reaches a price vector p in $(p^4,p^*]$ with corresponding activity level y_2 such that $\tilde{z}_3(p,(0,y_2)^T) = 0$ while besides $\tilde{z}_1(p,(0,y_2)^T) = 0$ then, due to Walras' law, also $\tilde{z}_2(p,(0,y_2)^T) = 0$ and $(p,(0,y_2)^T)$ is an equilibrium. Otherwise, the process reaches the price vector p^* with corresponding activity level vector $\tilde{y} = (0,\tilde{y}_2)^T$ being such that $\tilde{z}_3(p^*,\tilde{y}) = 0$ whereas $\tilde{z}_1(p^*,\tilde{y}) < 0$ and $\tilde{z}_2(p^*,\tilde{y}) > 0$. At p^* also activity 1 makes zero profit, so that its activity level y_1 is increased. Since commodity 3 is an input for activity 2 whereas it is an output of activity 1, an increase in y_1 must be matched by an increase in y_2 to keep commodity 3 in equilibrium. Lemma 5.2.1 then guarantees that y_1 eventually reaches a level y_1^* with corresponding y_2^* such that commodity 1 or commodity 2 becomes in equilibrium. But then all markets are in equilibrium and $(p^*,(y_1^*,y_2^*)^T)$ is an equilibrium.

5.4. Examples

In this section we apply the adjustment process to two specific examples. In both examples we consider an economy with three goods and two activities.

Example 5.4.1. We assume that the consumers can be represented by one consumer having a Cobb-Douglas utility function $u(x_1,x_2,x_3) =$

$x_1^{1/10}x_2^{2/5}x_3^{1/2}$, with x_i the amount of good i consumed. The initial endowments are $w = (1,1,1)^T$. Straightforward calculations yield that the consumer excess demand at price p equals $z(p) = ((1/10p_1) - 1, (2/5p_2) - 1, (1/2p_3) - 1)^T$. The activity vectors are given by $a^1 = (-3/2,1,1)^T$ and $a^2 = (-1,-77/27,11/9)^T$. This economy is depicted in Figure 5.4.1 with $[v,f] = \{p|p \cdot a^1 = 0\}$ and $\{\ell,r] = \{p|p \cdot a^2 = 0\}$.

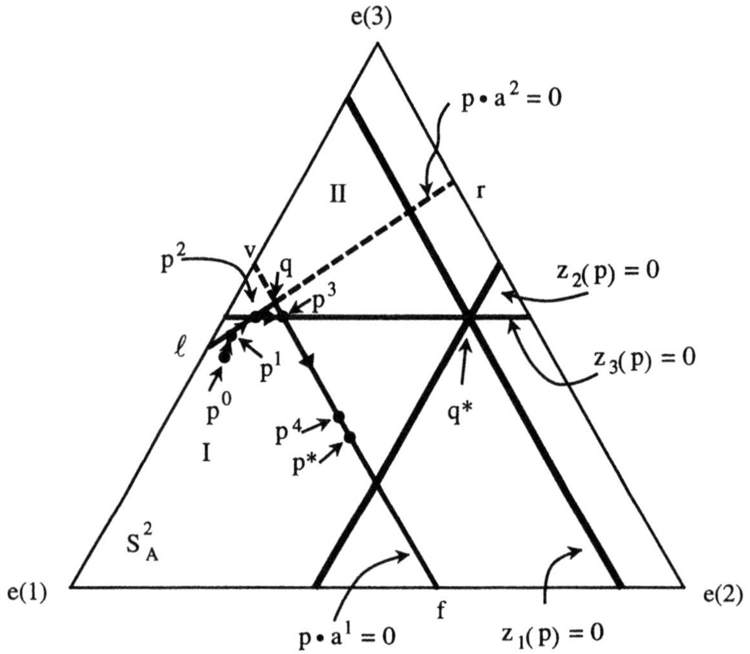

Figure 5.4.1. The set S^2_A for the economy of Example 5.4.1. For a price vector p in region I holds that $\text{sgn}(z(p)) = (-1,+1,+1)^T$. For region II, $\text{sgn}(z(p)) = (-1,+1,-1)^T$.

Observe that $\{p|z_j(p) = 0\} = \{p \in S^2|p_j = \alpha_j\}$ with $\alpha_1 = 1/10$, $\alpha_2 = 2/5$, $\alpha_3 = 1/2$ the budget shares of the consumer. We remark that d is not continuous on $\text{bd}(S^n)$. Therefore the curves $\{p|z_j(p) = 0\}$ and $\{p|z_h(p) = 0\}$ do not meet in $\{p|p_i = 0, i \neq j,h\}$. However an equilibrium for this economy exists because the conditions of Eaves [1987] are met. Another approach would be to cut off the boundary of the price space. This because near $\text{bd}(S^n)$ the demand for at least one good approaches infinity. Our process never reaches this boundary because on the path a very small price can only occur in case of a negative excess demand.

Again, the equilibrium price vector $q^* = (1/10,2/5,1/2)^\top$ of the corresponding pure exchange economy lies outside the set S^2_A being the convex hull of the vectors ℓ, q, f and e(1). Thus, in the economy with production there is no equilibrium with zero activity levels. We consider how the process reaches an equilibrium when starting from $(p^0,\underline{0})$ with $p^0 = (1/2,1/20,9/20)^\top$. The excess demand vector at p^0 is equal to $z(p^0) = (-4/5,7,1/9)^\top$. Thus, initially commodity 1 is in excess supply, whereas the goods 2 and 3 are in excess demand. Besides, activity 1 makes a loss of 1/4 whereas activity 2 makes a loss of 5/54. The process leaves $(p^0,\underline{0})$ by proportionally increasing the prices of goods 2 and 3, decreasing the price of good 1, and keeping both activity levels equal to zero. In the price set the process goes from p^0 into the direction opposite of e(1). At $(p^1,\underline{0})$, with $p^1 = (22/49,27/490,243/490)^\top$, the profit of activity 2 becomes zero. Then the activity level y_2 of activity 2 is increased until at $y_2 = 2/297$ commodity 3 becomes in equilibrium, i.e. $\tilde{z}_3(p^1,(0,2/297)^\top) = 0$. Then the process continues by keeping commodity 3 in equilibrium and the profit of activity 2 equal to zero, whereas the price ratio of commodity 3 varies between the minimum price ratio p_1/p_1^0 and the maximum price ratio p_2/p_2^0. In the figure the prices go towards r. At $p^2 = (11/25,3/50,1/2)^\top$, $z_3(p^2) = 0$ and hence the level of activity 2 needed to keep commodity 3 in equilibrium becomes zero. Then the process continues by keeping y_2 equal to zero and commodity 3 still in equilibrium. Activity 2 is allowed to become in a deficit situation. Thus, the process generates points $(p,\underline{0})$ with p going from p^2 towards q^* such that $\mathrm{sgn}(\tilde{z}(p,\underline{0})) = \mathrm{sgn}(z(p)) = (-1,+1,0)^\top$ and $p\cdot a^i < 0$, $i \in \{1,2\}$. At $(p^3,\underline{0})$ with $p^3 = (2/5,1/10,1/2)^\top$, activity 1 makes zero profit. Then the process increases y_1. Without simultaneously changing the prices this would give an excess supply on the market for commodity 3. Thus, the process has to adjust the prices such that the consumer excess demand for commodity 3 becomes positive. Projected on the price set, the process goes from p^3 towards f, i.e. it generates price vectors p at which $p\cdot a^1 = 0$ and $z_3(p) > 0$. At $p^4 = (2/5,6/25,9/25)^\top$ the corresponding activity level vector is $y = (7/18,0)^\top$ with $\tilde{z}(p^4,y) = (-1/6,5/18,0)^\top$. Observe that at p^4 the price ratio of commodity 3 is equal to the minimum price ratio, i.e. $p_3^4/p_3^0 = p_1^4/p_1^0 = 4/5$. Then, the adjustment process keeps both price ratios equal and minimal while bringing commodity 3 in excess supply by a further increase of y_1. At $y_1 = 1/2$ commodity 1 becomes in equilibrium ($\tilde{z}(p^4,(1/2,0)^{\top}) = (0,1/6,-1/9)^{\top}$). Thus, at p^4, the adaptation of the activity level changes $\mathrm{sgn}(\tilde{z})$ from $(-1,+1,0)^{\top}$ into $(0,+1,-1)^{\top}$. Then the process continues by adjusting the prices and y_1 such as to keep commodity 1 in equilibrium, while the profit of activity 1 is still equal to zero. The price ratio of good 1 is allowed to vary between the minimal price ratio p_3/p_3^0 and the maximal price ratio p_2/p_2^0. In the figure the process moves from p^4 towards f. At $p^* = (2/5,4/15,1/3)^{\top}$ the corresponding activity level vector $y^* = (1/2,0)^{\top}$ is such that $\tilde{z}(p^*,y^*) = \underline{0}$. Hence (p^*,y^*) is an equilibrium of this economy.

<u>Example 5.4.2</u>. In this example the excess demand function is based on Scarf [1960] and is given by $z(p) = (p_2-p_3,p_3-p_1,p_1-p_2)^{\top}$. The activities are $a^1 = (-3/2,1,1)^{\top}$ and $a^2 = (-1,-1,3/2)^{\top}$. All information is graphically presented in Figure 5.4.2. The unique equilibrium price vector of the related pure exchange economy is $q^* = (1/3,1/3,1/3)^{\top}$. Also in this example q^* lies outside S_A^2 being the convex hull of the vectors ℓ,p^*,f and e(1).

Now, let the process start at $(p^0,\underline{0})$, with $p^0 = (1/2,2/5,1/10)^{\top}$. The excess demand at that price vector equals $z(p^0) = (3/10,-2/5,1/10)^{\top}$. Thus, commodity 2 is in excess supply while the goods 1 and 3 are in excess demand. The process leaves $(p^0,\underline{0})$ by proportionally increasing the prices of the commodities 1 and 3, whereas the price of 2 is decreased. The activity levels remain zero because the process generates prices in $\mathrm{int}(S_A^2)$, i.e. the profits are negative. Projected on the price set, the process moves from p^0 in the direction opposite of e(2) (see Figure 5.4.2).

At $(p^1,\underline{0})$, with $p^1 = (5/7,1/7,1/7)^{\top}$, the market of good 1 becomes in equilibrium ($z(p^1) = (0,-4/7,4/7)^{\top}$). Now, the process adapts the prices such that the market of good 1 remains in equilibrium while the price ratio of good 1 is allowed to vary between the minimum price ratio p_2/p_2^0 and the maximum price ratio p_3/p_3^0. In the figure the process goes from p^1 into the direction of q^*. At $p^2 = (2/5,3/10,3/10)^{\top}$ where $z(p^2) = (0,-1/10,1/10)^{\top}$, activity 1 makes zero profit so that its activity level y_1 is increased from zero. However, an increase of y_1 at constant prices immediately yields an excess demand situation on the market for good 1 ($\tilde{z}_1(p^2,(y_1,0)^\top) = z_1(p^2)-y_1.a_1^1 = 0+y_1 3/2 > 0$). Thus, to keep good 1 in

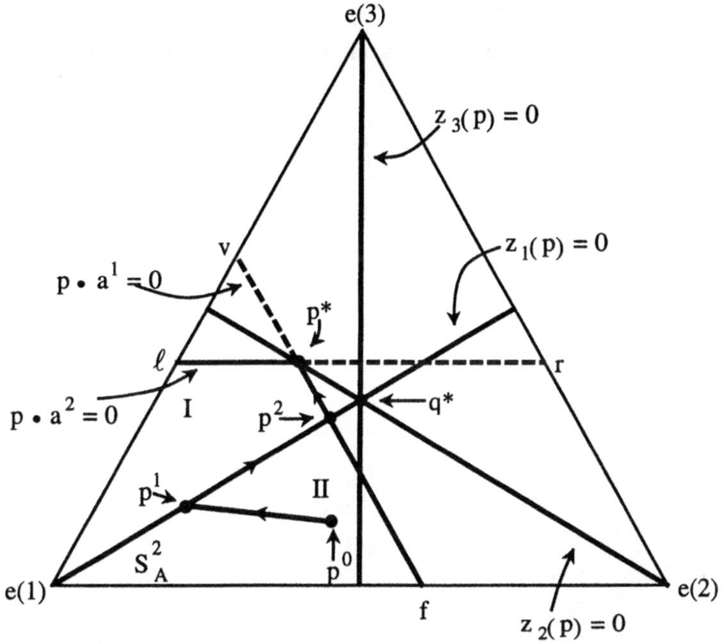

Figure 5.4.2. The set S_A^2 for the economy of Example 5.4.2. For a price vector in region I, II the sign of the vector $z(p)$ equals $(-1,-1,+1)^\top$, $(+1,-1,+1)^\top$, respectively.

equilibrium the prices have to be adjusted simultaneously making the consumers' excess demand for good 1 negative. Thus, projected on S_A^2 the process moves from p^2 towards v keeping the profit of activity 1 equal to zero. At $p^* = (2/5,1/5,2/5)^\top$ the consumers excess demand $z(p^*)$ equals $(-1/5,0,1/5)^\top$ and the level $\hat{y}_1$ making $\tilde{z}_1(p^*,(\hat{y}_1,0)^\top) = 0$ is equal to 2/15. It holds that $\tilde{z}(p^*,(\hat{y}_1,0)^\top) = (0,-2/15,1/15)^\top$. Besides, at p^* the profit level of activity 2 becomes zero so that the activity level y_2 is increased from zero. Because good 1 is an input for both activities, an increase in y_2 must be met by a decrease in y_1 to keep commodity 1 in equilibrium. Furthermore, it cannot happen that y_1 becomes zero. This because then y_2 must be equal to 1/5 in order to keep good 1 in equilibrium whereas

$\tilde{z}(p^*,(0,1/5)^T) = (0,1/5,-1/10)^T$. Thus, when going from $y = (2/15,0)^T$ to $y = (0,1/5)^T$ the market for good 2 turns from an excess supply situation into an excess demand. The opposite occurs for good 3. Thus, the increase in y_2 leads to a decrease in y_1 but before y_1 becomes zero the market for good 2 becomes in equilibrium at $y^* = (2/25,2/25)^T$. But then also good 3 must be in equilibrium and (p^*,y^*) is an equilibrium of this economy.

CHAPTER 6

Finding Nash equilibria in noncooperative games

In this chapter we consider the problem of finding Nash equilibria in mixed strategies for noncooperative games, in which the payoffs are listed in matrices or more generally in tensors. The main part of this chapter deals with bi-matrix games. Such a game is a noncooperative two-person game with a finite set of actions for each player. The Nash equilibrium is the standard equilibrium concept for a noncooperative game. It states that a strategy is an equilibrium when no player can improve upon his situation by deviating from his strategy while all other players keep on playing their strategies.

The standard method for solving a bi-matrix game is the method of Lemke and Howson [1964]. That method finds a Nash equilibrium by solving a related linear complementarity problem (LCP). This LCP is not defined on the space of strategies but on a nonnegative orthant. Each solution of the LCP uniquely determines an equilibrium on the strategy space.

The method proposed in this chapter finds a Nash equilibrium by solving a stationary point problem (SPP) on the strategy space. Thus, contrary to the Lemke-Howson procedure it directly operates on the strategy space. Both methods are complementary pivoting algorithms and therefore find under some nondegeneracy assumption a negatively indexed equilibrium (see Shapley [1974]). However, because the choice of the starting vector for the Lemke-Howson procedure is restricted, in case of multiple solutions that method may fail to reach certain negatively indexed equilibria. In our method the choice of the starting vector is free and therefore in principle we can find all negatively indexed equilibria by a repeated application of the algorithm from different starting vectors. We also show how the algorithm can find positively indexed equilibria by restarting it from negatively indexed equilibria already being found. Our method can be seen as a strategy adjustment process having an appealing game-theoretic interpretation. In this respect it is interesting that the algorithm always finds a perfect equilibrium whenever the starting vector lies in the interior of the strategy space, i.e. when at the start all actions are played with some positive probability.

There is also a third complementary pivoting procedure for solving a bi-matrix game, namely the Rhomb-Path method, which is due to Bastian and informally discussed in Todd [1978]. This method is more general than the Lemke-Howson method and will be discussed briefly further on in this chapter (Section 6.5). But also that method suffers from the same drawbacks as the Lemke-Howson method does. We remark that there are also algorithms that find all Nash equilibria with certainty. We mention the algorithms of Vorob'ev [1958], Kuhn [1961], Mangasarian [1964] and Winkels [1979]. Their methods are in principle equivalent. First the set of equilibria is characterized by a number of convex compact sets. The latter sets are localized by a procedure for finding extreme points of such sets. These methods do not have a game-theoretic interpretation and cannot be generalized to solve games with more than two players. How to generalize our algorithm to these games is illustrated at the end of this chapter.

The organization of the chapter is as follows. In Section 6.1 we show that the set of Nash equilibria in a bi-matrix game can be seen as the solution set of a stationary point problem. Furthermore, we give some intuition about how the algorithm operates. The formal steps of the procedure are presented in Section 6.2. Besides, in that section we state the nondegeneracy conditions under which the algorithm converges and we prove the perfectness of the equilibrium found by the procedure when starting from a strategy vector in the interior of the strategy space. In Section 6.3 we give some examples and provide a game-theoretic interpretation of the method. In Section 6.4 we show how the algorithm can be utilized to find positively indexed equilibria. In Section 6.5 we give a game-theoretic interpretation of the Lemke-Howson method and show how this method can be extended to find all equilibria in a bi-matrix game with one player having only two actions. Finally, Section 6.6 deals with the generalization of our method to games with more than two players.

The Sections 6.2, 6.3 and 6.4 are based on van den Elzen and Talman [1991], Section 6.5 on van den Elzen [1990], and Section 6.6 on van den Elzen and Talman [1992].

6.1. Solving the bi-matrix game as a stationary point problem

A noncooperative bi-matrix game is a game between two persons, called players, each having a finite number of actions and a payoff matrix indicating the payoff to that player when a certain pair of actions is played. Therefore, a bi-matrix game is a tuple (n_1,n_2,A,B), where $n_1(n_2)$ denotes the number of actions of player 1(2), while A(B) is the $(n_1 \times n_2)$-payoff matrix of player 1(2). More precisely, an element $a_{hi}(b_{hi})$ of the matrix A(B) gives the payoff to player 1(2) when player 1 plays action h and player 2 plays action i. Action k of player j is indicated by (j,k), $j \in \{1,2\}$. A bi-matrix game in which player 1(2) has $n_1(n_2)$ actions is often called an $(n_1 \times n_2)$ bi-matrix game.

A mixed strategy of player j, $j \in \{1,2\}$, is represented by a vector $x_j = (x_{j1},\ldots,x_{jn_j})^T$ in S^{n_j-1}. The number x_{jk}, $k \in \{1,\ldots,n_j\}$, is then the probability with which player j plays his k-th action at strategy x_j. Since the components of x_j are all nonnegative and sum up to one, x_j is indeed a vector of probabilities. We call S^{n_j-1} the strategy space of player j. In case actions are played with probability one we speak about pure strategies. They correspond to the vertices of the strategy space of the player.

The strategy space of the game is the simplotope S obtained by taking the Cartesian product of the strategy spaces of both players, i.e. $S = S^{n_1-1} \times S^{n_2-1}$. An element $x = (x_1,x_2)$ in S denotes a strategy vector of the game with x_j the strategy played by player j. The vertices of S correspond to the strategy vectors at which both players play a pure strategy.

The expected payoff to player 1(2) at strategy vector x equals $x_1 \cdot Ax_2$ $(x_2 \cdot B^T x_1)$. At a Nash equilibrium (NE) no player can improve upon his situation by unilaterally deviating from his strategy.

Definition 6.1.1. A *Nash equilibrium* of the bi-matrix game (n_1,n_2,A,B) is a strategy vector $x^* = (x_1^*,x_2^*)$ at which $x_1^* \cdot Ax_2^* = \max\{x_1 \cdot Ax_2^* | x_1 \in S^{n_1-1}\}$ and $x_2^* \cdot B^T x_1^* = \max\{x_2 \cdot B^T x_1^* | x_2 \in S^{n_2-1}\}$.

We now show that the set of Nash equilibria is equal to the set of solutions to a stationary point problem (SPP) on S with respect to some function $z : S \to \mathbb{R}^{n_1} \times \mathbb{R}^{n_2}$. For each x in S the vector $z(x) = (z_1(x), z_2(x))$, with $z_1(x) \in \mathbb{R}^{n_1}$ and $z_2(x) \in \mathbb{R}^{n_2}$, is given by

$$z_1(x) = Ax_2 \text{ and } z_2(x) = B^T x_1. \tag{6.1.1}$$

We call z the marginal payoff function. A component $z_{jk}(x)$ denotes the payoff at x to player j when he plays his k-th action while the other player i plays strategy x_i. Therefore, the expected payoff to player j at x can be written as $x_j \cdot z_j(x)$. Thus, x* is an NE if and only if

$$x_j \cdot z_j(x^*) \leqq x_j^* \cdot z_j(x^*) \quad , x_j \in S^{n_j - 1}, \; j \in \{1,2\}. \tag{6.1.2}$$

Therefore the problem of finding a Nash equilibrium is equivalent to finding a stationary point of z on S as defined in Definition 2.2.1. This follows from the proof of Theorem 2.2.4 and the subsequent remark. Because of the linearity of $x_j \cdot z_j(x^*)$ in x_j we only have to check (6.1.2) for the vertices of $S^{n_j - 1}$, $j \in \{1,2\}$. Thus, x* is an NE iff $z_{jk}(x^*) \leqq x_j^* \cdot z_j(x^*)$ for all $(j,k) \in I$. From this it is straightforward to derive that x* is an NE if and only if

$$z_{jk}(x^*) = \max_h z_{jh}(x^*) \text{ when } x_{jk}^* > 0, \; (j,k) \in I. \tag{6.1.3}$$

We say that an action k of player j is optimal at strategy x when $z_{jk}(x) = \max_h z_{jh}(x)$ and that at x player j is in equilibrium if at x all nonoptimal actions of j are played with probability zero. The interpretation of (6.1.3) is that at a Nash equilibrium the actions being played with positive probability are all optimal and hence that all players are in equilibrium.

The algorithm searches for a Nash equilibrium by solving the SPP as given in (6.1.3). Starting from an arbitrarily chosen strategy vector $v = (v_1, v_2)$ in S, the algorithm generates a piecewise linear path of points in S leading from v to a Nash equilibrium. More precisely, points

$x = (x_1,x_2) \in S$ on the path generated by the algorithm satisfy the following conditions. For $k \in \{1,\dots,n_1\}$,

$$\begin{aligned} x_{1k} &= b(x,v)v_{1k} && \text{if } z_{1k}(x) < \max_h z_{1h}(x) \\ x_{1k} &\geqq b(x,v)v_{1k} && \text{if } z_{1k}(x) = \max_h z_{1h}(x), \end{aligned}$$

and for $k \in \{1,\dots,n_2\}$, (6.1.4)

$$\begin{aligned} x_{2k} &= b(x,v)v_{2k} && \text{if } z_{2k}(x) < \max_h z_{2h}(x) \\ x_{2k} &\geqq b(x,v)v_{2k} && \text{if } z_{2k}(x) = \max_h z_{2h}(x), \end{aligned}$$

where $0 \leqq b(x,v) := \min_{(j,h)}\{x_{jh}/v_{jh} | v_{jh} > 0\} \leqq 1$.

Observe that $x = v$ satisfies (6.1.4) with $b(x,v) = 1$. Also each Nash equilibrium x^* satisfies (6.1.4) with $b(x^*,v) = 0$ or with $v_{jk} = 0$ for all (j,k) for which $z_{jk}(x^*) < \max_h z_{jh}(x^*)$. In the latter case the nonoptimal actions are already played with probability zero at the start. In both cases, $x^*_{jk} = b(x^*,v).v_{jk}$ is equal to zero when at x^* action (j,k) is not optimal. Under some nondegeneracy condition the set of points satisfying (6.1.4) contains a piecewise linear path, P, from v to a Nash equilibrium. This path will be followed by the algorithm. The notion of nondegeneracy will be made precise further on, but it is for example required that at $x = v$ both $\max_k z_{1k}(x)$ and $\max_\ell z_{2\ell}(x)$ are attained for a unique index. Thus, at the starting strategy vector v each player has only one optimal action. Suppose these maxima are attained for the actions $(1,k_1)$ and $(2,k_2)$, respectively. Clearly, from v, along P, $b(x,v)$ must decrease from 1. Thus, according to (6.1.4), initially vectors x in S are generated such that all the x_{1k}, $k \neq k_1$, and x_{2k}, $k \neq k_2$, are relatively decreased ($x_{ik} = b(x,v).v_{ik}$ for $(i,k) \notin \{(1,k_1),(2,k_2)\}$), while both x_{1k_1} and x_{2k_2} are increased in order to keep x_1 in S^{n_1-1} and x_2 in S^{n_2-1}. This is continued till either $b(x,v)$ becomes 0 and a Nash equilibrium is reached, or a strategy vector x is reached at which for some j, $z_{jk}(x) = z_{jk_j}(x)$, $k \neq k_j$. Then x_{jk} is also relatively increased. In general, the algorithm generates strategy vectors x such that all the x_{jk}/v_{jk}, related to the indices (j,k) for which $z_{jk}(x) < \max_h z_{jh}(x)$, are minimal, i.e. equal to $b(x,v)$ ($x_{jk} = 0$

if $v_{jk} = 0$). As soon as one of these components of $z(x)$, say $z_{j\ell}(x)$, becomes equal to $\max_h z_{jh}(x)$, then $x_{j\ell}/v_{j\ell}$ is increased from $b(x,v)$ ($x_{j\ell}$ is increased from zero if $v_{j\ell} = 0$), while keeping $z_{j\ell}(x)$ maximal for j. On the other hand, if a vector x is generated such that x_{jr}/v_{jr} for some (j,r) with $z_{jr}(x) = \max_h z_{jh}(x)$ becomes minimal, i.e. equal to $b(x,v)$ (x_{jr} becomes 0 if $v_{jr} = 0$), then vectors y are generated with y_{jr} equal to $b(y,v).v_{jr}$ while $z_{jr}(y)$ is decreased from $\max_h z_{jh}(y)$.

To illustrate the foregoing we conclude this section with an example.

Example 6.1.1. Consider the bi-matrix game (n_1,n_2,A,B) with $n_1 = n_2 = 2$ and

$$A = \begin{bmatrix} 4 & 0 \\ -1 & 3 \end{bmatrix}, \quad B = \begin{bmatrix} 4 & -3 \\ 2 & 4 \end{bmatrix}.$$

In this game each player has two actions. If player 1 plays action 2 and player 2 action 1 then player 1 gets a payoff of -1 and player 2 a payoff of 2. The marginal payoff function z_1 of player 1 is given by $z_1(x) = (4x_{21},-x_{21}+3x_{22})^\top$. Similarly, $z_2(x) = (4x_{11}+2x_{12},-3x_{11}+4x_{12})^\top$. This game has three Nash equilibria: $((1,0)^\top,(1,0)^\top)$, $((0,1)^\top,(0,1)^\top)$, and $((2/9,7/9)^\top,(3/8,5/8)^\top)$. The first two equilibria are Nash equilibria in pure strategies. For ease of notation we denote in the remaining of this chapter a vector $x = ((x_{11},x_{12})^\top,(x_{21},x_{22})^\top)$ by $((x_{11},x_{12}),(x_{21},x_{22}))$. The strategy space for this game is equal to $S = S^1 \times S^1 = \{x \in \mathbb{R}^2_+ \times \mathbb{R}^2_+ \mid x_{11}+x_{12} = 1,\ x_{21}+x_{22} = 1\}$, and is displayed in Figure 6.1.1.

Let the starting point v be given by $v = ((1/8,7/8),(1/2,1/2))$. It is straightforward to verify that $z(v) = ((2,1),(9/4,25/8))$. Thus at v, $z_{11}(v)$ is maximal for $j = 1$ and $z_{22}(v)$ for $j = 2$. So, the algorithm leaves v in the direction of $((1,0),(0,1))$, i.e. x_{11} and x_{22} are both increased. The algorithm continues in this way till the vector $a = ((2/9,7/9),(4/9,5/9))$ is reached at which $z(a) = ((16/9,11/9),(22/9,22/9))$, i.e. at $x = a$, $z_{21}(x)$ has become equal to $z_{22}(x)$. Observe that along the segment $[v,a]$ the number $b(x,v)$ decreases from 1 to 8/9. Next, the algorithm

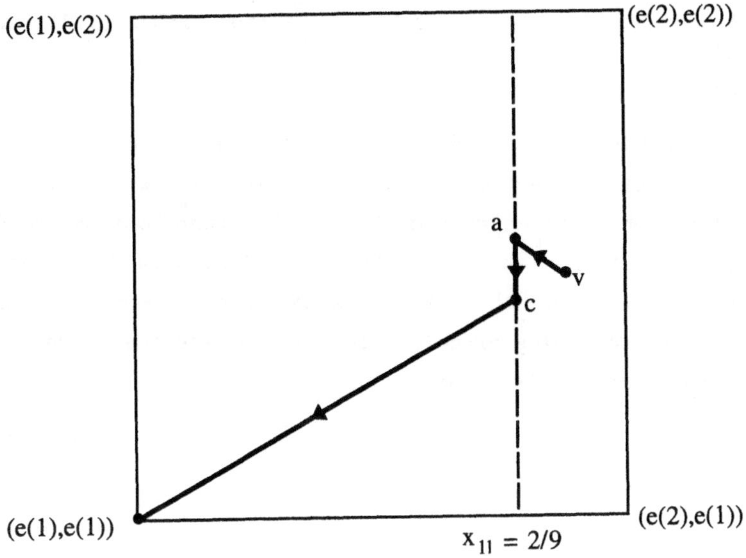

Figure 6.1.1. The path P of points satisfying (6.1.4) from v to a Nash equilibrium.

continues, according to (6.1.4), by increasing x_{21} relatively away from x_{12} while keeping $z_{21}(x)$ equal to $z_{22}(x)$. It is easy to verify that the latter holds along the line segment [a,c]. In fact, $z_{21}(x) = z_{22}(x) = 22/9$ if $x_{11} = 2/9$. At $c = ((2/9,7/9),(5/9,4/9))$ with $z(c) = ((20/9,7/9), (22/9,22/9))$ we still have $z_{21}(c) = z_{22}(c)$ and $z_{11}(c) > z_{12}(c)$. But observe that at x = c, x_{22}/v_{22} has become equal to $b(x,v) = x_{12}/v_{12} = 8/9$. When keeping $z_{22}(x)$ equal to $z_{21}(x)$ beyond x = c one would have to generate points x with $x_{22}/v_{22} < x_{12}/v_{12}$ which violates (6.1.4). In that case one would leave S at the vector ((2/9,7/9),(1,0)). But instead the algorithm continues from x = c by keeping x_{22} relatively minimal, i.e. x_{22}/v_{22} is kept equal to x_{12}/v_{12}, while, according to (6.1.4), $z_{22}(x)$ is decreased from $\max_h z_{2h}(x) = z_{21}(x)$. In this way the algorithm reaches the vertex

((1,0),(1,0)) which is a pure Nash equilibrium with z((1,0),(1,0)) = ((4,-1),(4,-3)).

The idea behind the procedure is to generate a sequence of vectors x along which the set T of actions (j,k) for which $z_{jk}(x) = \max_h z_{jh}(x)$ grows while the probabilities related to all the other actions are driven down to zero, since if they are zero a Nash equilibrium is found as follows from (6.1.3). However, the set T does not need to grow monotonically (cf. point c in the foregoing example). This guarantees the convergence of the algorithm to a Nash equilibrium.

6.2. The procedure

The algorithm is a complementary pivoting procedure. To implement it we first rewrite (6.1.4) into a system of linear equations. By substituting (6.1.1) we obtain that the process generates from v the (piecewise linear) path P of strategy vectors $x \in S$ satisfying for $k \in \{1,\dots,n_1\}$,

$$\begin{aligned} x_{1k} &= bv_{1k} && \text{if } A^k x_2 < \beta_1 \\ x_{1k} &\geqq bv_{1k} && \text{if } A^k x_2 = \beta_1, \end{aligned}$$

and for $k \in \{1,\dots,n_2\}$ (6.2.1)

$$\begin{aligned} x_{2k} &= bv_{2k} && \text{if } x_1 \cdot B_k < \beta_2 \\ x_{2k} &\geqq bv_{2k} && \text{if } x_1 \cdot B_k = \beta_2, \end{aligned}$$

where $b := b(x,v)$, C^k and C_k respectively being the k-th row and k-th column of a matrix C and where $\beta_1 = \max_k A^k x_2$, and $\beta_2 = \max_k x_1 \cdot B_k$.

For each x in S satisfying (6.2.1) there is at least one set $T \subset I$ such that $x_{jk} \geqq bv_{jk}$ and $z_{jk}(x) = \beta_j$ for all $(j,k) \in T$ while $z_{ih}(x) \leqq \beta_i$ and $x_{ih} = bv_{ih}$ for all $(i,h) \notin T$. From this observation we obtain that the procedure generates for a sequence of subsets T of I with $T_j := T \cap I(j) \neq \emptyset$, starting from x = v, strategy vectors x in S such that

$$x_{1k} = bv_{1k} \quad \text{and} \quad A^k x_2 \leqq \beta_1 \quad \text{if } (1,k) \notin T$$
$$x_{1k} \geqq bv_{1k} \quad \text{and} \quad A^k x_2 = \beta_1 \quad \text{if } (1,k) \in T,$$

and (6.2.2)

$$x_{2k} = bv_{2k} \quad \text{and} \quad x_1 \cdot B_k \leqq \beta_2 \quad \text{if } (2,k) \notin T$$
$$x_{2k} \geqq bv_{2k} \quad \text{and} \quad x_1 \cdot B_k = \beta_2 \quad \text{if } (2,k) \in T.$$

We now show how the path P can be followed by a sequence of linear programming (ℓ.p.) pivot steps in a system of linear equations obtained from (6.2.2). Denote the set of points x in S satisfying (6.2.2) for given $T \subset I$ by B(T). Assuming nondegeneracy as given below, each nonempty B(T) is a line segment in S. Thus, the algorithm reaches a Nash equilibrium via a (finite) sequence of sets B(T), $T \subset I$. We need only one ℓ.p. step to traverse such a B(T). The system of linear equations in which the ℓ.p. pivot step is made is obtained from (6.2.2) by introducing slack variables for each inequality. The slack variables for the inequalities $A^h x_2 \leqq \beta_1$, $(1,h) \notin T$, are denoted by μ_{1h} and those for $x_1 \cdot B_h \leqq \beta_2$, $(2,h) \notin T$, by μ_{2h}. The slack variables for the inequalities $x_{jk} \geqq bv_{jk}$, $(j,k) \in T$, are denoted by λ_{jk}. Adding these slacks to (6.2.2) we obtain that x belongs to B(T) iff there exists $0 \leqq b \leqq 1$ and for $j \in \{1,2\}$ there exist $\lambda_{jk} \geqq 0$ for $(j,k) \in T$, $\mu_{jk} \geqq 0$ for $(j,k) \notin T$, $\beta_j \in \mathbb{R}$ such that for $j \in \{1,2\}$

$$x_j = bv_j + \Sigma_{(j,k)\in T_j} \lambda_{jk} e^{n_j}(k) \text{ with } \Sigma_{(j,k)\in T_j} \lambda_{jk} = 1-b, \qquad (6.2.3)$$

while

$$Ax_2 + \Sigma_{(1,h)\notin T}\, \mu_{1h} e^{n_1}(h) = e^{n_1}\beta_1$$

and (6.2.4)

$$B^T x_1 + \Sigma_{(2,h)\notin T}\, \mu_{2h} e^{n_2}(h) = e^{n_2}\beta_2,$$

where $T_j := T \cap I(j)$.

In the sequel of this chapter we use the following simplifying notation. The vector $e^{n_j}(h)$, $j \in \{1,2\}$, is denoted by $\bar{e}^j(h)$ and similarly e^{n_j} by $\bar{e}^j$.

Substituting (6.2.3) in (6.2.4) gives the system of linear equations

$$b\begin{bmatrix} Av_2 \\ B^{T}v_1 \\ 1 \\ 1 \end{bmatrix} + \sum_{(1,k)\in T} \lambda_{1k}\begin{bmatrix} \underline{0} \\ B_k^{T} \\ 1 \\ 0 \end{bmatrix} + \sum_{(2,k)\in T} \lambda_{2k}\begin{bmatrix} A_k \\ \underline{0} \\ 0 \\ 1 \end{bmatrix} + \sum_{(1,h)\notin T} \mu_{1h}\begin{bmatrix} \bar{e}^1(h) \\ \underline{0} \\ 0 \\ 0 \end{bmatrix}$$

$$+ \sum_{(2,h)\notin T} \mu_{2h}\begin{bmatrix} \underline{0} \\ \bar{e}^2(h) \\ 0 \\ 0 \end{bmatrix} - \beta_1\begin{bmatrix} \bar{e}^1 \\ \underline{0} \\ 0 \\ 0 \end{bmatrix} - \beta_2\begin{bmatrix} \underline{0} \\ \bar{e}^2 \\ 0 \\ 0 \end{bmatrix} = \begin{bmatrix} \underline{0} \\ \underline{0} \\ 1 \\ 1 \end{bmatrix}, \qquad (6.2.5)$$

where B_k^{T} denotes the k-th column of the matrix B^{T}.

In the system (6.2.5) there are n_1+n_2+2 equations and n_1+n_2+3 variables. The last two equations of (6.2.5) reflect the property that $\sum_{(j,k)\in T_j}\lambda_{jk} = 1-b$, $j \in \{1,2\}$. These equations can be eliminated by substituting for $j \in \{1,2\}$ one of the λ_{jk}'s, say λ_{jk_j}, by

$$\lambda_{jk_j} = 1 - b - \sum_{\substack{(j,k)\in T_j \\ k\neq k_j}} \lambda_{jk}. \qquad (6.2.6)$$

Let T_j^1, $j \in \{1,2\}$, be the set of indices defined by $T_j^1 = T_j \backslash \{(j,k_j)\}$, and let $T^1 = T_1^1 \cup T_2^1$. All of this together and the substitution of b by 1-b', gives the following system of equations

$$b'\begin{bmatrix} A_{k_2} - Av_2 \\ B_{k_1}^{T} - B^{T}v_1 \end{bmatrix} + \sum_{(1,k)\in T^1} \lambda_{1k}\begin{bmatrix} \underline{0} \\ B_k^{T} - B_{k_1}^{T} \end{bmatrix} + \sum_{(2,k)\in T^1} \lambda_{2k}\begin{bmatrix} A_k - A_{k_2} \\ \underline{0} \end{bmatrix}$$

$$+ \sum_{(1,h)\notin T} \mu_{1h}\begin{bmatrix} \bar{e}^1(h) \\ \underline{0} \end{bmatrix} + \sum_{(2,h)\notin T} \mu_{2h}\begin{bmatrix} \underline{0} \\ \bar{e}^2(h) \end{bmatrix} - \beta_1\begin{bmatrix} \bar{e}^1 \\ \underline{0} \end{bmatrix}$$

$$- \beta_2\begin{bmatrix} \underline{0} \\ \bar{e}^2 \end{bmatrix} = \begin{bmatrix} -Av_2 \\ -B^{T}v_1 \end{bmatrix}. \qquad (6.2.7)$$

The latter system has only n_1+n_2 equations and n_1+n_2+1 variables. Its variables must satisfy $\lambda_{jk} \geqq 0$ for $(j,k) \in T^1$, $\sum_{(j,k)\in T^1_j} \lambda_{jk} \leqq b'$ for $j \in \{1,2\}$, $0 \leqq b' \leqq 1$, $\mu_{ih} \geqq 0$ for $(i,h) \notin T$. A solution to this system is denoted by $(b',\lambda_1,\lambda_2,\mu_1,\mu_2,\beta_1,\beta_2)$ and corresponds to a point $x = (x_1,x_2)$ in $B(T)$ as defined in (6.2.3). An ℓ.p. pivot step of the algorithm is made in system (6.2.7).

Before stating the formal steps of the algorithm we give a non-degeneracy condition that guarantees the convergence of the procedure. This condition is explained in more detail after the description of the algorithm.

Assumption 6.2.1. (nondegeneracy assumption). At each solution $(b',\lambda_1,\lambda_2,\mu_1,\mu_2,\beta_1,\beta_2)$ of (6.2.7) at most one of the constraints $0 \leqq b' \leqq 1$, $\lambda_{jk} \geqq 0$ for $(j,k) \in T^1$, $b' \geqq \Sigma_{(j,k)\in T^1_j}\lambda_{jk}$, $\mu_{ih} \geqq 0$ for $(i,h) \notin T$, is binding, unless $b' = 1$ or $v_{ih} = 0$ for all $(i,h) \notin T$.

Step 0 [Initialization].

Choose an arbitrary vector v in S. If v is a Nash equilibrium then the algorithm stops. Else calculate the (unique) indices $(1,k_1)$ and $(2,k_2)$ for which $A^{k_1}v_2 = \max_k A^k v_2$ and $v_1 \cdot B_{k_2} = \max_k v_1 \cdot B_k$. Furthermore, set $T^1 = \emptyset$, $b' = 0$, $\beta_1 = A^{k_1}v_2$, $\beta_2 = v_1 \cdot B_{k_2}$, $\mu_{1h} = \beta_1 - A^h v_2$ for $h \neq k_1$, and $\mu_{2h} = \beta_2 - v_1^T B_h$ for $h \neq k_2$. Increase b' from 0 in (6.2.7) and go to Step 1.

Step 1.

a. If b' becomes 1 then let the solution of (6.2.7) be $(1,\lambda_1^*,\lambda_2^*,\mu_1^*,\mu_2^*,\beta_1^*,\beta_2^*)$. The vector $x^* = (x_1^*,x_2^*)$, with $x_j^* = \Sigma_{(j,k)\in T_j}\lambda_{jk}^*\bar{e}^j(k)$, $j \in \{1,2\}$, is a Nash equilibrium and the algorithm stops.

b. If λ_{jk} becomes 0 for some $(j,k) \in T^1$ then T^1 becomes $T^1\backslash\{(j,k)\}$ and go to Step 2a.

c. If $\Sigma_{(j,k)\in T^1_j} \lambda_{jk}$ becomes equal to b' for some $j \in \{1,2\}$ then, according to (6.2.6), λ_{jk_j} becomes 0. Go to Step 2b.

d. If μ_{ih} becomes zero for some $(i,h) \notin T$ then go to Step 3.

Step 2.

a. Increase the complementary variable μ_{jk} from zero by pivoting into system (6.2.7) the column $(\bar{e}^1(k)^\top, \underline{0}^\top)^\top$ if j = 1, and $(\underline{0}^\top, \bar{e}^2(k)^\top)^\top$ if j = 2. Return to Step 1.

b. Substitute the largest λ_{jk}, say $\lambda_{j\ell}$, by $b' - \Sigma_{(j,h)\in T_j^1,\ h\neq k_j,\ell} \lambda_{jh}$. Increase μ_{jk_j} from zero by pivoting the related column into system (6.2.7). T^1 becomes $T^1 \backslash \{(j,\ell)\}$, k_j becomes ℓ, and return to Step 1.

Step 3.

If additionally $v_{jk} = 0$ for all $(j,k) \notin T \cup \{(i,h)\}$ then let the solution be $(b'^*, \lambda_1^*, \lambda_2^*, \mu_1^*, \mu_2^*, \beta_1^*, \beta_2^*)$. The vector $x^* = (x_1^*, x_2^*)$, with $x_j^* = (1-b'^*)v_j + \Sigma_{(j,k)\in T_j} \lambda_{jk}^* \bar{e}^j(k)$, $j \in \{1,2\}$, is a Nash equilibrium and the algorithm stops. Else increase the complementary variable λ_{ih_1} from zero by pivoting its related column into system (6.2.7). T^1 becomes $T^1 \cup \{(i,h)\}$ and return to Step 1.

Let us make a few remarks. The algorithm starts with increasing b' from zero. From (6.2.6) we derive that this implies that both λ_{1k_1} and λ_{2k_2} are increased from zero. In Step 1c, the variable λ_{jk_j} becomes zero. Then we have to adapt system (6.2.7) by eliminating another λ_{jk} to take over the role of λ_{jk_j}. In principle any λ_{jk}, $(j,k) \in T_j^1$, can be taken. We suggest to take the largest one, say $\lambda_{j\ell}$. This substitution can easily be performed in (6.2.7) by adding the column related to $\lambda_{j\ell}$ to the column related to b' and subtracting the same column from the columns related to λ_{jh}, $h \neq \ell$.

Assumption 6.2.1 is standard in linear programming and assures that all steps of the algorithm are unique. More precisely, when this nondegeneracy condition holds it cannot occur that more than one constraint in Step 1 becomes binding simultaneously. This is only allowed when the algorithm stops. Assumption 6.2.1 also guarantees the uniqueness of the indices $(1,k_1)$ and $(2,k_2)$ in Step 0. To see this, suppose that

$A^{k_1}v_2 = \max_k A^k v_2 = A^{\ell} v_2$ for some $\ell \neq k_1$. But then $\mu_{1\ell} = 0$ and two constraints (the other one concerns b' = 0) are binding. This is the only restriction on the choice of the starting vector.

The nondegeneracy condition guarantees the convergence of the algorithm. First, observe that the solution set of (6.2.7) is bounded for given T. Since $\beta_j = \max_k z_{jk}(x)$ and z is a linear function on the compact set S, β_j must be finite and therefore also the μ_{ih}'s with $(i,h) \notin T$. This together with Assumption 6.2.1 implies that for each T the solution set of (6.2.7) and hence also B(T) is either empty or a line segment with two end points. The algorithm starts by traversing $B(T^0)$, with $T^0 = \{(1,k_1), (2,k_2)\}$. If the algorithm operates in some B(T) and λ_{jk} becomes zero for some $(j,k) \in T$ then the algorithm continues in $B(T\backslash\{(j,k)\})$. Similarly, the algorithm continues in $B(T \cup \{(i,h)\})$ if μ_{ih} becomes zero for some $(i,h) \notin T$ and not a Nash equilibrium has been reached. Assumption 6.2.1 guarantees that these transitions are unique so that no cycling can occur. Because there are only a finite number of possible subsets T the algorithm must reach a Nash equilibrium within a finite number of ℓ.p. steps.

Our nondegeneracy condition is weaker than the nondegeneracy condition on games of Lemke-Howson [1964] (see also Shapley [1974]). We only need this condition to hold along the path generated by the algorithm. The last pivot step might be degenerated. In other words, it is possible to apply our procedure to games that are degenerated in the sense of Lemke-Howson. In the next section we give an example.

At the end of this section we discuss whether something can be said about game-theoretic properties holding for the equilibrium found by our algorithm. More concrete, we would like to know whether the equilibrium found is isolated, quasi-strong, regular, essential, perfect, or proper. The precise definition of these concepts can be found for example in van Damme [1983]. It turns out that our algorithm finds a perfect equilibrium whenever it starts from an interior strategy vector. This is interesting the more because for bi-matrix games an equilibrium is perfect iff it is an equilibrium in undominated strategies (see Theorem 3.2.2 in van Damme [1983]). Other properties may not hold for the equilibrium reached by our procedure. This is due to the fact that the algorithm can successfully be applied to games having not such equilibria.

Thus, the positive result is that our algorithm finds a perfect equilibrium whenever the starting point v is in the inner of the strategy space, i.e. when $v_{jk} > 0$ for all $(j,k) \in I$. To prove the statement it is most convenient to define a perfect equilibrium as a limit point of a sequence of ε-perfect equilibria.

<u>Definition 6.2.2</u>. For $\varepsilon > 0$, the strategy vector $x(\varepsilon) \in S$ is an *ε-perfect equilibrium* if it is completely mixed and satisfies for $j \in \{1,2\}$

$$\text{if } z_{jk}(x(\varepsilon)) < z_{j\ell}(x(\varepsilon)) \text{ then } x_{jk}(\varepsilon) \leqq \varepsilon, \ \forall k,\ell.$$

The strategy vector $x^* \in S$ is a *perfect equilibrium* if $x^* \in \lim_{\varepsilon\downarrow 0}\{x(\varepsilon)\}$, with for all $\varepsilon > 0$, $x(\varepsilon)$ an ε-perfect equilibrium.

Observe that an ε-perfect equilibrium does not need to be a Nash equilibrium. The concept only states that all actions are played with positive probability while nonoptimal actions are played with a very small probability. The perfectness concept gives a stability requirement to the equilibrium. In case players make small mistakes, i.e. they play nonoptimal strategies with a small but nonzero probability, an equilibrium is perfect if these mistakes give no incentive to any of the players to deviate from the equilibrium strategy. Another characterization of a perfect equilibrium x^* in a bi-matrix game is that an equilibrium is perfect if it is undominated. The latter means that each player j has no strategy $\bar{x}_j$ such that $\bar{x}_j$ is at least as good as x^*_j against all possible strategies of the other player while it is better against at least one strategy of the opponent. Finally, observe from Definition 6.2.2 that every completely mixed Nash equilibrium is perfect. This because then all actions are optimal and so for the sequence of ε-perfect equilibria we can take a repetition of the equilibrium itself. In the next section we illustrate the concept in an example.

<u>Theorem 6.2.3</u>. If the starting vector v lies in the interior of S then the algorithm finds a perfect Nash equilibrium.

Proof. From (6.1.4) we know that the process generates a path of strategy vectors x in S satisfying

$$\begin{aligned} x_{jk} &= b(x,v).v_{jk} \qquad \text{if } z_{jk}(x) < \max_{\ell} z_{j\ell}(x) \\ x_{jk} &\geqq b(x,v).v_{jk} \qquad \text{if } z_{jk}(x) = \max_{\ell} z_{j\ell}(x), \end{aligned} \tag{6.2.8}$$

where $0 \leqq b(x,v) = \min_{(i,h)}\{x_{ih}/v_{ih} | v_{ih} > 0\} \leqq 1$. We already argued that all Nash equilibria in the interior of S are perfect. Therefore, suppose the algorithm finds an NE x^* on the boundary. Then, because $v_{jk} > 0$ for all $(j,k) \in I$, $b(x,v)$ decreases in the last iteration down to zero. More precisely, $b(x,v)$ decreases along a line segment from say y to x^* from $b(y,v)$ to $b(x^*,v) = 0$. Let $I(x^*) := \{(j,k) \in I | x^*_{jk} = 0\}$ and let (r,ℓ) be an index for which $y_{r\ell} = \max_{(i,p)\in I(x^*)} y_{ip}$. It is easily checked from (6.2.8) that each vector x on $[y,x^*)$ is an ε-perfect equilibrium with $\varepsilon = x_{r\ell}$. The limit point x^* of the sequence $\{x(\varepsilon)\}_{\varepsilon\downarrow 0}$ is then perfect by definition.

□

Note that we proved above that the algorithm not only finds a perfect equilibrium when starting in the interior but also that the last linear piece generated consists of ε-perfect equilibria. We emphasize that the result above is not trivial. This because the algorithm can be applied to games which are degenerated in the sense of Lemke-Howson. For such games not all equilibria are perfect. However, our procedure succeeds to find one when being started in the interior of the strategy space. In Section 6.3 we illustrate this with an example.

6.3. Game-theoretic interpretation

In this section we want to explain how the algorithm works in terms of strategies and payoffs. Technically speaking the algorithm roughly operates as follows. From the start, the variable b' is increased from zero. Recall from (6.2.6) that an increase of b' means that both λ_{1k_1} and λ_{2k_2} are increased and that b is decreased (from 1). As soon as μ_{jk} for some $(j,k) \notin T$ becomes zero, its complementary variable λ_{jk} is increased from zero and vice versa. From (6.2.3) we deduce that a positive λ_{jk} means that the relative probability with which player j uses his k-th action is larger than the relative minimum ($x_{jk} > bv_{jk}$), whereas $\lambda_{jk} = 0$ indicates that $x_{jk} = bv_{jk}$. Similarly, from (6.2.4) we infer that $\mu_{jk} > 0$ corresponds to action (j,k) being nonoptimal for player j, i.e. $z_{jk}(x) < \max_h z_{jh}(x)$, whereas $\mu_{jk} = 0$ means that action (j,k) is optimal for player j. With all of this together it is straightforward to derive the game-theoretic interpretation of the adjustments made by the algorithm.

Initially the probabilities related to the unique optimal actions of both players are increased, whereas the probabilities related to all other actions are proportionally decreased. If the latter probabilities all become zero then a Nash equilibrium is reached. This because then all nonoptimal actions are played with zero probability. Else, the algorithm eventually generates a strategy vector at which for some player a second action becomes optimal. Then the procedure continues by keeping that action optimal whereas the corresponding probability is relatively increased. In general, the algorithm generates strategy vectors at which the nonoptimal actions are played with probabilities all being, relative to the starting probabilities, equal to each other and smaller than each probability with which an optimal action is played. As soon as a nonoptimal action becomes optimal, its relative probability is increased from the probabilities related to the nonoptimal actions. On the other hand, if a probability with which an optimal action is played becomes relatively equal to the probabilities of the nonoptimal actions then it is kept equal to those and the algorithm continues by making the related action nonoptimal. When the probabilities of the nonoptimal actions for all players are zero a Nash equilibrium has been found.

For a more specific illustration we apply the algorithm to the example presented earlier (see Figure 6.1.1). For the solution $(b',\mu_1,\mu_2,\beta_1,\beta_2)$ of (6.2.7) at v it holds that $\beta_1 = (Av_2)_1 = 2$, $\beta_2 = (B^{\mathsf{T}}v_1)_2 = 25/8$, $\mu_{12} = \beta_1-(Av_2)_2 = 2-1 = 1$, $\mu_{21} = \beta_2-(B^{\mathsf{T}}v_1)_1 = 25/8-9/4 = 7/8$, while $b' = 0$. The algorithm leaves v by increasing b' from zero, i.e. λ_{11} and λ_{22} are increased (Step 0). At the vector a, the variable μ_{21} has become zero (Step 1d) whereas $b' = 1/9$, $\beta_1 = (Aa_2)_1 = 16/9$, $\beta_2 = (B^{\mathsf{T}}a_1)_2 =$

22/9, μ_{12} = 5/9. Thus, the algorithm continues from a by increasing λ_{21} from zero (Step 3). The solution at c is $\beta_1 = (Ac_2)_1 = 20/9$, $\beta_2 = (B^T c_1)_1 = (B^T c_1)_2 = 22/9$, $\mu_{12} = \beta_1 - (Ac_2)_2 = 13/9$, $\lambda_{21} = b' = 1/9$. Thus, λ_{22} has become zero (Step 1c). In system (6.2.7), $(2,k_2)$ becomes (2,1) and the algorithm continues by increasing μ_{22} from zero (Step 2b). In the next step the algorithm reaches the Nash equilibrium ((1,0),(1,0)) at which b' = 1 (Step 1a).

The game-theoretic interpretation of the adjustments along the latter path is as follows. At v, action (1,1) is optimal for player 1 and action (2,2) is optimal for player 2. Now the algorithm increases from v the probabilities with which these actions are played and decreases with the same rate the probabilities of all other actions. The algorithm continues in this way till it generates the strategy vector a at which player 2 becomes in equilibrium ($z_{21}(a) = z_{22}(a)$). From a the algorithm generates strategy vectors x by relatively increasing probability x_{21} away from x_{12} while keeping player 2 in equilibrium. At strategy vector c the probability with which player 2 plays his second action has become relatively equal to the probability related to the only nonoptimal action (1,2). Then the algorithm distorts the equilibrium situation of player 2. It continues by generating vectors x at which for player 2 the action (2,2) is nonoptimal, i.e. $z_{22}(x)$ is made smaller than $z_{21}(x)$. Meanwhile, the probabilities x_{12} and x_{22} are kept relatively equal to each other but smaller than x_{11} and x_{21}. In this way the Nash equilibrium ((1,0),(1,0)) is reached.

We conclude this section with an application of our algorithm to a bi-matrix game which is degenerated in the sense of Lemke-Howson [1964] (see also Shapley [1974]). The algorithm can not only be successfully applied to this game but it also finds a perfect Nash equilibrium whenever it starts in the interior, although there are an infinite number of equilibria not being perfect. The game we consider is the bi-matrix game with payoff matrices

$$A = \begin{bmatrix} 4 & 1 \\ 0 & 1 \end{bmatrix} \quad \text{and} \quad B = \begin{bmatrix} 2 & 1 \\ 1 & 2 \end{bmatrix}.$$

This game is graphically represented in Figure 6.3.1. The piece-wise linear curve heavily drawn denotes the set of optimal strategies of player 1 against player 2, also called the best reply set R^1 of player 1. We see that player 1 plays his first action with probability one against all strategies of player 2, except when player 2 plays his second action with probability one. In that case player 1 is indifferent between his actions. Similarly, one can derive the best reply set R^2 of player 2, indicated in the figure by the dashed piecewise linear curve. The Nash equilibria coincide with the intersections of R^1 and R^2. In the figure {(1,1),(2,1)} indicates that in the corresponding subset of S the actions (1,1) and (2,1) are optimal. Similarly, for {(1,1),(2,2)}.

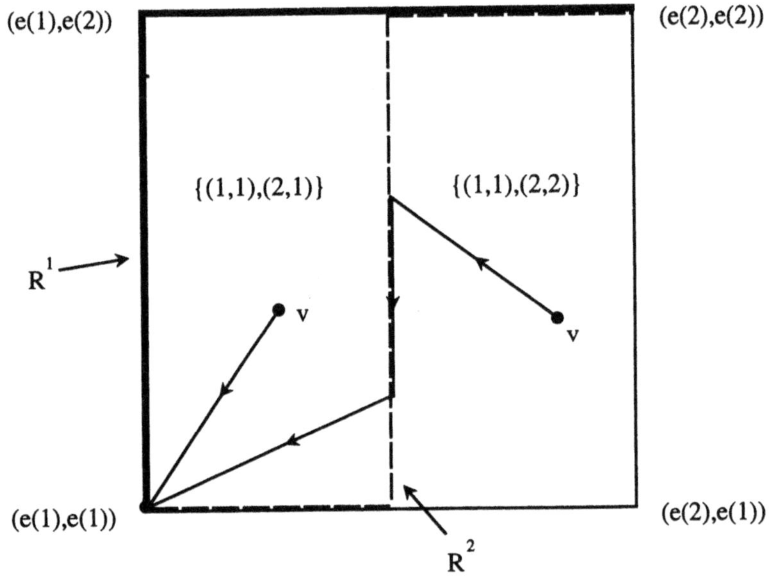

Figure 6.3.1. The set of Nash equilibria consists of the unique perfect Nash equilibrium ((1,0),(1,0)) and the set $\{x \in S \mid x_{11} \leq 1/2,\ x_2 = (0,1)\}$.

When the algorithm starts in the interior of S it always finds the perfect equilibrium ((1,0),(1,0)). We have given two possible paths in the figure. Observe that all Nash equilibria except ((1,0),(1,0)) fail to be perfect. Consider for example the equilibrium ((0,1),(0,1)). If player 2 makes a small mistake and plays his first action with arbitrary small probability then player 1 immediately plays his first action. Hence, this equilibrium is unstable against small mistakes.

6.4. How to find more equilibria

As already mentioned, like other pivoting procedures, the algorithm always finds a negatively indexed equilibrium, where the sign of the index is the sign of the determinant of a certain matrix related to that equilibrium. For more details we refer to Shapley [1974]. Equilibria can have positive or negative indices. In general, a bi-matrix game possesses an odd number of isolated equilibria of which the number of equilibria being negatively indexed is one more than the number of equilibria with positive index. To become more precise concerning the generality of the statement, observe that an $(n_1 \times n_2)$ bi-matrix game can be viewed upon as an element in $\mathbb{R}^{2n_1 . n_2}$. Then, the set of bi-matrix games satisfying the statement above is dense and open in $\mathbb{R}^{2n_1 . n_2}$. It is for this large set that the following is of interest.

Again consider (6.1.4) for a given bi-matrix game. The set of strategy vectors obeying (6.1.4) is determined by the specific starting point v. Let us therefore denote the set of such vectors by $\mathcal{B}_v$. We already argued that $\mathcal{B}_v$ contains a piecewise linear path connecting v and an NE. However, in general this set is larger. When a bi-matrix game has an odd number of isolated equilibria then $\mathcal{B}_v$ contains besides the path connecting v and a negatively indexed NE, a set of disjoint piecewise linear loops and paths, each path connecting a positively indexed and a negatively indexed equilibrium. We provide some intuition for this fact. Recall that for each strategy vector x obeying (6.1.4) there is a set $T \subset I$ such that x satisfies (6.2.2). Let B(T) be the set of all strategy vectors satisfying (6.2.2) for a given $T \subset I$. As in Chapter 3, B(T) is the intersection of subsets A(T) and C(T) defined as $A(T) = \emptyset$ when $v_{jk} = 0$ for all $(j,k) \notin T$, while otherwise

$$A(T) = \{x \in S | x_{jk} = b(x,v).v_{jk},\ (j,k) \notin T\}$$

and

$$C(T) = \{x \in S | z_{jk}(x) = \max_h z_{jh}(x),\ (j,k) \in T\}.$$

The dimension of a nonempty A(T) equals $|T|-1$ whereas generically either $C(T) = \emptyset$ or $\dim(C(T)) = n_1+n_2-|T|$. Thus, for $T \subset I$, either B(T) is empty or its dimension is equal to 1. It is straightforward to show now that $\mathcal{B}_v$, being the union of all the subsets B(T), is a set of 1-dimensional paths and loops. Each path and loop is piecewise linear because of the linearity of z. Clearly, the end points of a path not being equal to v are Nash equilibria according to (6.1.4). Moreover, each linear piece of a path or loop corresponds to some B(T) and can therefore be generated by a pivot step in (6.2.7). Finally, the end points of a path connecting two Nash equilibria are differently indexed. As an illustration serves Figure 6.4.1 in which for the game of Example 6.1.1 the set $\mathcal{B}_v$ has been drawn with v = ((1/8,7/8),(1/2,1/2)) (see also Figure 6.1.1).

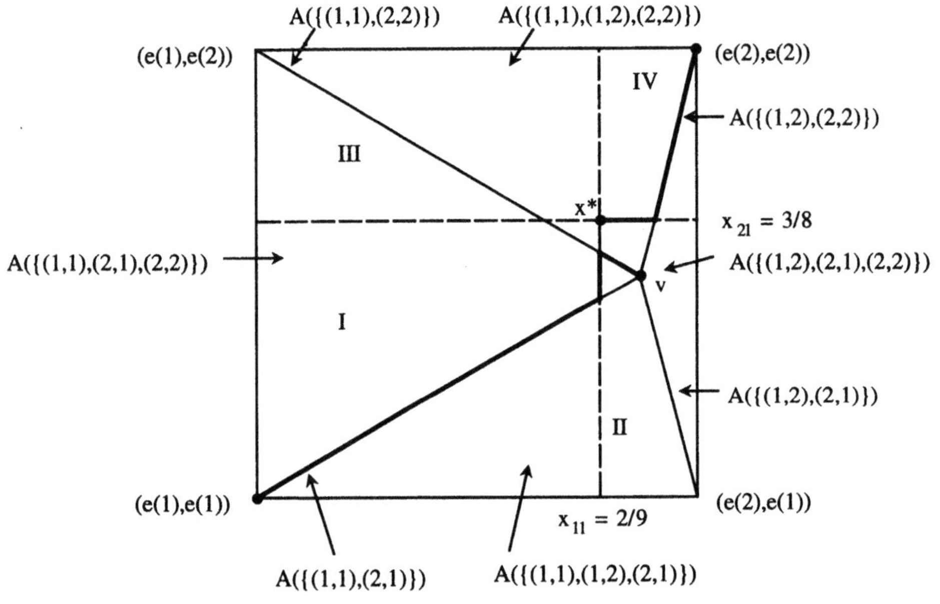

<u>Figure 6.4.1</u>. The set $\mathcal{B}_v$ consists of a piecewise linear path connecting v and the negatively indexed equilibrium ((1,0),(1,0)) and a piecewise linear path linking the negatively indexed NE ((0,1),(0,1)) and the positively indexed NE x*. The regions I, II, III, IV denote C({(1,1),(2,1)}), C({(1,1),(2,2)}), C({(1,2),(2,1)}), C({(1,2),(2,2)}), respectively.

In this figure we also depict all sets A(T) and C(T). Observe that the subsets A(T), $|T| = 3$, constitute a subdivision of S. We remark that

Lemke and Howson [1964] already proved the existence of a complementary path between a starting vector and an NE and of paths connecting pairs of Nash equilibria. However, as we will see in the next section, the Lemke-Howson procedure may not be able to reach certain paths.

Now we are ready to explain how to find more equilibria with the procedure. First we apply the algorithm several times, each time starting from a different starting vector. In this way we may find several Nash equilibria, say k. Then we pick a starting vector, say v, from which we found one of these equilibria. Next, we initialize the algorithm at each of the other k-1 equilibria. Each time the algorithm then traces a piece-wise linear path in $\mathcal{B}_v$ from a negatively indexed to a positively indexed equilibrium. In this way we find 2k-1 equilibria. We illustrate this along with the Example 6.1.1. To follow the description below let us turn to Figure 6.4.2, where we again depict the set $\mathcal{B}_v$ related to v = ((1/8,7/8), (1/2,1/2)) and also the path leading from the starting vector $\bar{v}$ to the NE ((0,1),(0,1)).

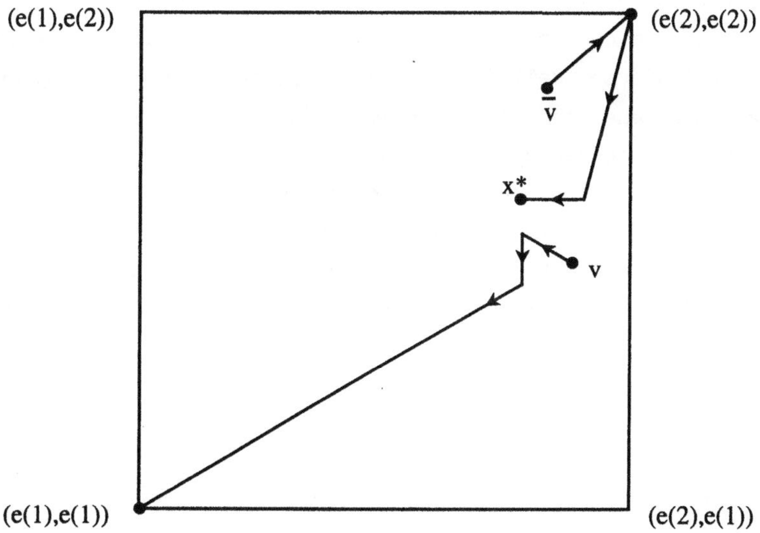

Figure 6.4.2. The set $\mathcal{B}_v$ and the path of the algorithm starting from $\bar{v}$.

To find all three Nash equilibria we first apply the algorithm from v and find ((1,0),(1,0)). Next, we try other starting vectors till we find one from which we reach another NE. For example we find $\bar{v}$ with related equilibrium ((0,1),(0,1)). To find the third, positively oriented, NE we restart the algorithm in ((0,1),(0,1)) and traverse in $\mathcal{B}_v$ the piecewise linear path from ((0,1),(0,1)) to x^* in two pivoting steps. To carry out the latter we have to substitute v for $\bar{v}$ in the final linear system related to the algorithm having been started at $\bar{v}$. We leave ((0,1),(0,1)) by decreasing b' from 1. That the algorithm can be applied is due to the fact that all vectors on the piecewise linear path connecting x^* and ((0,1),(0,1)) satisfy (6.1.4) and therefore lie in B(T) for $T \subset I$.

Recall that the algorithm converges to a negatively indexed NE when starting from almost any strategy vector. We also stated that almost any bi-matrix game possesses an odd number of equilibria. Besides, these equilibria are perfect. Now, to each of the negatively indexed equilibria corresponds a, relative to S, open set of strategy vectors in the sense that whenever the algorithm starts from a vector in this set the related equilibrium is reached. The set is called the attraction set of that equilibrium. The sets are open relative to S because a small distortion of the starting vector leads to small (continuous) changes in the path generated by the algorithm and the equilibrium reached is the same. Because the algorithm only fails to reach an equilibrium when starting from a vector lying in some lower dimensional set, the closures of the attraction sets form a subdivision of the strategy space. The faces of an attraction set are polyhedra because of the linearity of z and the fact that the subsets A(T) form a subdivision. In Figure 6.4.3 this is illustrated for Example 6.1.1.

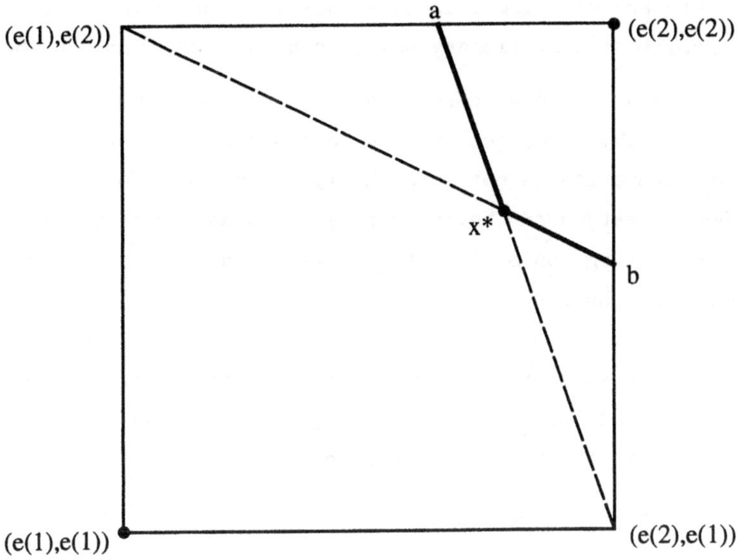

Figure 6.4.3. The two attraction sets belonging to the negatively indexed equilibria. The set below the segments $[a,x^*]$ and $[x^*,b]$ is the set corresponding to the equilibrium ((1,0),(1,0)) and above to ((0,1),(0,1)).

Observe that when in the example the algorithm is started from a vector lying in the intersection of the closures of the attraction sets, i.e. on $[a,x^*]$ or $[x^*,b]$, the positively indexed equilibrium x^* is reached.

To conclude this section we repeat that our algorithm may find more than one equilibrium. However, to deduce when we have found all equilibria we need some additional theory on topological degree. We return to this point at the end of the next section.

6.5. Interpretation and generalization of the Lemke-Howson algorithm

The standard method for finding an NE in a bi-matrix game (n_1,n_2,A,B) is the procedure of Lemke-Howson [1964]. Like our method it is a complementary pivoting procedure. However, our method solves the equilibrium problem as a stationary point problem (SPP) on S whereas Lemke-Howson solves a linear complementarity problem (LCP) on $\mathbb{R}_+^{n_1}\times\mathbb{R}_+^{n_2}$. Because the latter method does not operate on the strategy space its game-theoretic interpretation is not immediately clear. Therefore we discuss how the Lemke-Howson method can be interpreted also as a strategy adjustment procedure operating on S. From this interpretation the differences with our method become obvious.

Let us start reviewing the Lemke-Howson method (L-H). This exposition is based on Garcia and Zangwill [1981]. As already indicated, L-H finds an NE in the bi-matrix game (n_1,n_2,A,B) by solving a related LCP on $\mathbb{R}_+^{n_1}\times\mathbb{R}_+^{n_2}$. More precisely, it searches for a vector $y = (y_1,y_2) \in \mathbb{R}_+^{n_1}\times\mathbb{R}_+^{n_2}$ such that

$$\begin{bmatrix} w_1 \\ w_2 \end{bmatrix} := \begin{bmatrix} -\bar{e}^1 \\ -\bar{e}^2 \end{bmatrix} + \begin{bmatrix} 0 & -A \\ -B^T & 0 \end{bmatrix}\begin{bmatrix} y_1 \\ y_2 \end{bmatrix} \geqq 0,\ w_1\cdot y_1 = 0,\ w_2\cdot y_2 = 0. \tag{6.5.1}$$

It is assumed that $A,B < 0$ in (6.5.1). This can be done without loss of generality because decreasing each payoff element with the same amount does not change the equilibria. Now, the set of Nash equilibria corresponds to the set of solutions to (6.5.1). More concretely, $\hat{y} = (\hat{y}_1,\hat{y}_2)$ solves the LCP (6.5.1) if and only if $\bar{x} = (\bar{x}_1,\bar{x}_2)$ with $\bar{x}_j = (\bar{e}^j\cdot\hat{y}_j)^{-1}\hat{y}_j$, $j \in \{1,2\}$, is an NE (see Garcia and Zangwill [1981], ch. 19). First observe that the LCP above is equivalent to the problem of finding a vector $y = (y_1,y_2) \in \mathbb{R}_+^{n_1}\times\mathbb{R}_+^{n_2}$ such that

$$F_{jk}(y) := \min\{y_{jk},w_{jk}\} = 0, \qquad (j,k) \in I. \tag{6.5.2}$$

We call (y_{jk},w_{jk}) a pair of complementary variables. Now, the method generates a path of vectors y in $\mathbb{R}_+^{n_1}\times\mathbb{R}_+^{n_2}$ satisfying $F_{11}(y) \geqq 0$ and $F_{jk}(y) = 0$, $(j,k) \neq (1,1)$, till it finds a vector $\hat{y}$ with $F_{11}(\hat{y}) = 0$. The starting vector is the vector $\tilde{y}$ for which $\tilde{y}_{11} = -1/b_{1s}$ (> 0), with $b_{1s} = \max_j b_{1j}$, and $\tilde{y}_{2s} = -1/a_{rs}$ (> 0), with $a_{rs} = \max_i a_{is}$, whereas all its other components are zero. Observe from (6.5.1) that at $\tilde{y}$ it holds that $w_{2s} = w_{1r} = 0$ while $w_{jk} \geqq 0$ for $(j,k) \notin \{(2,s),(1,r)\}$. Thus, $\tilde{y}$ solves (6.5.2) if and only if $r = 1$. If not, the method starts by increasing y_{1r} away from zero, whereas the other zero variables are kept zero. Notice that at $\tilde{y}$, (y_{1r},w_{1r}) is the only pair of complementary variables that are both zero and observe from (6.5.1) that an increase of y_{1r} affects w_2 and y_{11}. The latter because w_{2s} is kept equal to zero. Now, two things can occur. If y_{11} becomes zero, the algorithm stops with a solution $\hat{y}$ at which $\hat{y}_{1r} = -1/b_{rs}$ and $\hat{y}_{2s} = -1/a_{rs}$. Else a variable w_{2k}, $k \neq s$, becomes zero. Then the method continues by increasing its complementary variable y_{2k} from zero. In general, if $y_{jk}(w_{jk})$ becomes zero then $w_{jk}(y_{jk})$ is increased from zero. The algorithm stops as soon as y_{11} or w_{11} becomes zero.

We remark that along the path traced by the Lemke-Howson procedure there is exactly one pair of complementary variables that are both nonzero, namely (y_{11},w_{11}). In fact, this corresponds to the fact that the L-H method searches for an equilibrium by tracing a path along the edges of a certain polytope related to the payoff matrices. The Rhomb-Path method, which has already been mentioned in the introduction, is similar to the Lemke-Howson method but then there are at most two of such pairs. That method searches along two-dimensional faces of the same polytope. For more details we refer to Todd [1978].

Example 6.5.1. Consider the game (n_1,n_2,A,B) with $n_1 = 2$, $n_2 = 3$,

$$A = \begin{bmatrix} 1 & -2 & 5 \\ 3 & 2 & 4 \end{bmatrix}, \qquad B = \begin{bmatrix} -2 & 4 & 3 \\ 5 & -2 & 1 \end{bmatrix}.$$

To apply L-H we first subtract 6 from each matrix element to make A and B negative. System (6.5.1) then becomes

$$\begin{bmatrix} w_{11} \\ w_{12} \\ w_{21} \\ w_{22} \\ w_{23} \end{bmatrix} = \begin{bmatrix} -1 \\ -1 \\ -1 \\ -1 \\ -1 \end{bmatrix} + \begin{bmatrix} 0 & 0 & 5 & 8 & 1 \\ 0 & 0 & 3 & 4 & 2 \\ 8 & 1 & 0 & 0 & 0 \\ 2 & 8 & 0 & 0 & 0 \\ 3 & 5 & 0 & 0 & 0 \end{bmatrix} \begin{bmatrix} y_{11} \\ y_{12} \\ y_{21} \\ y_{22} \\ y_{23} \end{bmatrix} \geqq 0, \quad \begin{matrix} y_1, y_2 \geqq 0 \\ w_1 \cdot y_1 = 0, \; w_2 \cdot y_2 = 0. \end{matrix}$$

At the start we have that $y_{11} = 1/2$, $s = 2$, $y_{22} = 1/4$, and $r = 2$. Then the algorithm increases y_{12} from 0 while keeping $w_{22} = 0$. It is easily verified that w_{23} becomes zero when y_{12} equals 1/14 ($y_{11} = 3/14$). Then the method continues by increasing y_{23} from zero. This leads to a decrease of y_{22} because w_{12} is kept equal to zero. If y_{23} becomes 1/3 then y_{22} equals 1/12 and w_{11} equals zero. L-H stops with the vector $\hat{y} = ((3/14,1/14),(0,1/12,1/3))$. The corresponding NE equals $x^* = ((3/4,1/4),(0,1/5,4/5))$.

We now give a game-theoretic interpretation of the path generated by the Lemke-Howson algorithm. We do this by transforming the vectors y and w into strategy vectors and marginal payoffs. Again consider (6.5.1). From this system we deduce that $w_{1k} = 0$ ($w_{2k} = 0$) if and only if $(Ay_2)_k = \max_j (Ay_2)_j$ ($(B^T y_1)_k = \max_j (B^T y_1)_j$). Recalling (6.5.1) we see that each vector y generated by L-H uniquely corresponds to the strategy vector $x = (x_1,x_2)$ with $x_j = (\bar{e}^j \cdot y_j)^{-1} y_j$, $j \in \{1,2\}$, and so $z_{jk}(x) = \max_\ell z_{j\ell}(x)$ if and only if $w_{jk} = 0$. In particular, the starting vector $\tilde{y}$ corresponds to the pure strategy vector at which player 1 (2) plays (1,1) ((2,s)), whereas action (1,r) is optimal for player 1 and (2,s) for player 2. Thus, L-H starts with player 1 playing his first action while player 2 plays his best reply against (1,1). This is action (2,s), with $b_{1s} = \max_j b_{1j}$. In the case that (1,1) is the best reply of player 1 against (2,s), the starting vector is also an NE. This corresponds to $r = 1$. If not, then L-H leaves the starting strategy vector by increasing from zero the probability with which player 1 plays his best reply to (2,s). As soon as some action (j,k) becomes optimal (w_{jk} becomes zero) then the corresponding probability is increased (y_{jk} is increased). On the other hand, if the probability with which an optimal action (j,k) is played becomes zero (y_{jk} becomes zero), then the related action is made nonoptimal (w_{jk} is increased from zero).

The process stops as soon as action (1,1) becomes optimal or if the probability with which this action is played becomes zero. At that point all actions played are optimal (cf. (6.1.3)). We again illustrate this with the previous example.

Example 6.5.1 (continued). The strategy space $S = S^1 \times S^2$ for this game is given in Figure 6.5.1. Here, the rectangular CDEF represents the strategy vectors x at which for player 1 both his actions are optimal, i.e. $z_{11}(x) = z_{12}(x)$. These vectors are characterized by $x_{22} = -1+3x_{23}/2$. Above that plane action (1,1) is optimal for player 1 and below the plane action (1,2) is optimal for him. The bold piecewise linear curve connecting ((1,0),(0,1,0)) and $\hat{x}$ = ((0,1),(1,0,0)) denotes the best reply curve of player 2. A vector $x = (x_1,x_2)$ on that curve indicates that x_2 is a best reply against strategy x_1 of player 1. For player 2 action (2,2) is optimal if $x_{11} \geqq 3/4$, action (2,3) if $4/9 \leqq x_{11} \leqq 3/4$, whereas action (2,1) is optimal whenever $x_{11} \leqq 4/9$. Now, L-H starts with player 1 playing (1,1) and player 2 playing his best reply (2,2). The best reply of 1 against (2,2) is his action (1,2) and hence the probability with which 1 plays (1,2) is increased from 0 (x_{11} is decreased from 1). For player 2 action (2,2) is optimal to play till x_{12} becomes 1/4. At that vector player 2 becomes indifferent between playing (2,2) and (2,3). Then the probability with which he plays action (2,3), i.e. x_{23} is increased. Finally, at x^* = ((3/4,1/4),(0,1/5,4/5)) player 1 becomes in equilibrium and an NE is reached.

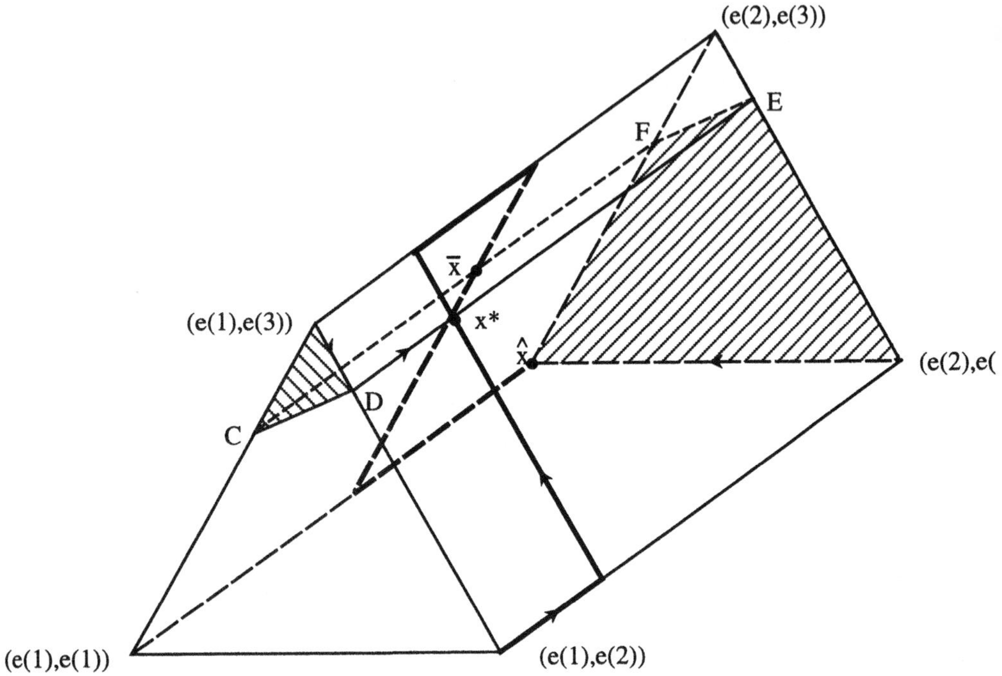

Figure 6.5.1. Illustration of the Lemke-Howson algorithm projected on the strategy space.

Note that we do not need to make the payoff matrices negative in order to apply Lemke-Howson on the strategy space. On S we only have to take account for optimal actions. In the original L-H algorithm the negativeness is needed for the choice of the starting vector. Because we have now put both our process and Lemke-Howson in a unifying framework we are able to compare them. But before doing that we first give a complementary pivoting algorithm by which we can generate the projected path of the Lemke-Howson procedure. We saw that the Lemke-Howson method in fact generates a path of strategy vectors $x \in S$, starting from v with $v_{11} = v_{2s} = 1$, characterized by

$$x_{11} \geqq 0 \text{ and } z_{11}(x) \leqq \max_{\ell} z_{1\ell}(x)$$
$$x_{jk} = 0 \text{ or } z_{jk}(x) = \max_{\ell} z_{j\ell}(x), \qquad (j,k) \neq (1,1). \tag{6.5.3}$$

Denoting the set of optimal actions by T and using the same notation as in (6.2.5) we obtain that each vector x on the path has to satisfy the system

$$x_{11}\begin{bmatrix} \underline{0} \\ B_1^T \\ 1 \\ 0 \end{bmatrix} + \Sigma_{(1,k)\in T}\, x_{1k}\begin{bmatrix} \underline{0} \\ B_k^T \\ 1 \\ 0 \end{bmatrix} + \Sigma_{(2,k)\in T}\, x_{2k}\begin{bmatrix} A_k \\ \underline{0} \\ 0 \\ 1 \end{bmatrix} + \Sigma_{(1,h)\notin T}\, \mu_{1h}\begin{bmatrix} \bar{e}^1(h) \\ \underline{0} \\ 0 \\ 0 \end{bmatrix}$$

$$+ \Sigma_{(2,h)\notin T}\, \mu_{2h}\begin{bmatrix} \underline{0} \\ \bar{e}^2(h) \\ 0 \\ 0 \end{bmatrix} - \beta_1\begin{bmatrix} \bar{e}^1 \\ \underline{0} \\ 0 \\ 0 \end{bmatrix} - \beta_2\begin{bmatrix} \underline{0} \\ \bar{e}^2 \\ 0 \\ 0 \end{bmatrix} = \begin{bmatrix} \underline{0} \\ \underline{0} \\ 1 \\ 1 \end{bmatrix}, \tag{6.5.4}$$

where all variables except β_1 and β_2 are nonnegative. A solution to (6.5.4) is denoted by $(x_1,x_2,\mu_1,\mu_2,\beta_1,\beta_2)$. The final system is then obtained by substituting $x_{11} = 1 - \Sigma_{k\neq 1} x_{1k}$ and $x_{2s} = 1 - \Sigma_{k\neq s} x_{2k}$. This system consists of n_1+n_2 equations with n_1+n_2+1 unknowns. At the start we have that $T = \{(2,s)\}$, $x_{11} = x_{2s} = 1$, $\beta_1 = z_{1r}(v)$, $\beta_2 = z_{2s}(v)$, and $\mu_{jk} = \beta_j - z_{jk}(v)$, $(j,k) \notin T$. If $\mu_{11} = 0$, then the starting vector v is an NE. Else $\mu_{1r} = 0$ for some $r \neq 1$ and x_{1r} is increased from zero, i.e. T becomes $\{(1,r),(2,s)\}$. In general, this algorithm operates similar to the one described in Section 6.2. If x_{jk} becomes zero then T becomes $T\backslash\{(j,k)\}$ and the algorithm continues by increasing μ_{jk} from zero. On the other hand, if μ_{jk} becomes zero then T becomes $T \cup \{(j,k)\}$ and x_{jk} is increased from zero. The algorithm stops whenever x_{11} or μ_{11} becomes zero. The case in which x_{2s} becomes zero is similar to the situation in which λ_{jk_j} becomes zero in the algorithm described in Section 6.2. For the algorithm to work we need a nondegeneracy condition similar to Condition 6.2.1. For example, at the start each player should have a unique optimal action. For completeness we also apply this algorithm to Example 6.5.1.

Example 6.5.1 (continued). At $v = ((1,0),(0,1,0))$ we have $T = \{(2,2)\}$, $x_{11} = x_{22} = 1$, $\beta_1 = z_{12}(v) = 2$, $\beta_2 = z_{22}(v) = 4$, $\mu_{11} = 4$, $\mu_{21} = 6$ and $\mu_{23} = 1$. All other variables are zero. The algorithm leaves v by increasing x_{12} from 0. If x_{12} becomes 1/4, μ_{23} becomes zero. Then the algorithm proceeds by increasing x_{23} from 0 till the NE $x^* = ((3/4,1/4), (0,1/5, 4/5))$ is reached at which μ_{11} becomes 0.

When we compare the Lemke-Howson procedure with our method we see some remarkable differences. First of all, the choice of the starting vector is arbitrary for our algorithm whereas the starting vector is constructed in the L-H method. Next, computational experiments give arguments for the idea that our method works fast for games of large dimension with both players having more or less an equal number of actions. L-H performs better for other games. The third and most important difference lies in the game-theoretic interpretation of the paths generated by both algorithms. Let us now discuss in more detail the points mentioned.

A very important feature of our procedure is the fact that the starting vector can be chosen arbitrarily. In the Lemke-Howson procedure the construction of the starting vector is part of the algorithm. It chooses a starting vector at which player 1 plays his first action whereas player 2 plays his best reply upon that action. By renumbering actions and players there are at most n_1+n_2 different starting vectors from which the L-H method can start. To illustrate this we again consider the game of Example 6.5.1.

Example 6.5.1 (continued). We applied L-H to this game when starting from the vector $((1,0),(0,1,0))$. Renumbering the strategies of player 1 gives the starting vector $((0,1),(1,0,0))$ which is an NE. By interchanging the players we obtain two more starting vectors, namely $((0,1),(0,1,0))$ and $((1,0),(0,0,1))$. Starting from the first one, L-H reaches $((0,1), (1,0,0)$ via one pivot step, whereas it reaches $((3/4,1/4),(0,1/5,4/5))$ in two pivot steps when it starts from the second one. All possible paths are illustrated in Figure 6.5.1.

Because L-H can only start from a limited number of strategy vectors it can occur that this method cannot find certain negatively indexed Nash equilibria. Consider for example the game given in Shapley [1974]. This concerns a bi-matrix game in which both players have three actions with payoff matrices

$$A = \begin{bmatrix} 0 & 3 & 0 \\ 2 & 2 & 0 \\ 3 & 0 & 1 \end{bmatrix} \quad \text{and} \quad B = \begin{bmatrix} 0 & 2 & 3 \\ 3 & 2 & 0 \\ 0 & 0 & 1 \end{bmatrix}.$$

To solve this game L-H can start in only three vectors, ((1,0,0),(0,0,1)), ((0,1,0),(1,0,0)), and ((0,0,1),(0,0,1)). Note that interchanging the players gives no additional starting vectors because the payoff structures of the players are identical. Starting from these vectors, L-H always finds the NE ((0,0,1),(0,0,1)). However, this game possesses a second negatively indexed equilibrium, $x^* = ((1/3,2/3,0),(1/3,2/3,0))$. Our algorithm finds this equilibrium if it starts for example in $v = ((0,1,0),(0,1,0))$. At that vector the actions (1,1) and (2,1) are optimal for the two players. By increasing the related probabilities, x^* is reached in one step. Note that the probabilities related to actions (1,3) and (2,3) remain zero. We refer to the article of Todd [1978] for more details about the structure of this kind of examples. In that article the structure of games that give problems for the Rhomb-Path method is also sketched.

When dealing with algorithms, an important issue is of course the computational speed. Therefore we did some computational experiments with both algorithms. For that purpose we considered bi-matrix games of different dimension with randomly generated payoff elements and determined the number of pivot steps needed to reach an NE. Of course, for our algorithm this number in principle depends upon the starting vector selected. For that reason we applied our algorithm three times to each game. First, we started from the barycentre, being the vector at which each player plays all actions with equal probability. Next, we took as starting vector a vector at which both players play a rationalizable strategy (see Bernheim [1984] and Pearce [1984]). At such a vector only actions are played that are a best reply against an action played with positive probability by the other player. Finally, we considered the case in which our procedure starts from the vector at which both players only play their first action. The latter choice represents the case in which the starting vector is chosen without prior knowledge. It turns out that our algorithm performs better for games of relatively large size in which the players have more or less an equal number of actions. L-H performs better in other cases. Concerning the choice of the starting vector it makes overall seen no difference whether one starts from the barycentre or from a rationalizable strategy vector. However, starting from an arbitrary pure strategy vector gives definitely worse results.

The major differences between both algorithms concern the game-theoretic interpretation of the paths they generate. As can be seen from (6.5.3), L-H generates from the starting vector a path of strategy vectors at which player 2 is in equilibrium, i.e. he plays optimal against the strategy of player 1. In other words, L-H operates on the best reply set of player 2 which is defined by

$$R^2 = \{(x_1,x_2) \in S \mid x_2 \cdot B^T x_1 \geqq \bar{x}_2 \cdot B^T x_1,\ \bar{x}_2 \in S^{n_2-1}\}$$

The strategies of player 1 are such that all actions played with positive probabilities except (1,1) are optimal. Now, L-H generates the piecewise linear curve in R^2 originating at $v_1 = \bar{e}^1(1)$ by first increasing the probability with which player 1 plays his optimal action against the reply of player 2. In our procedure both players are at the start typically in disequilibrium. Throughout the algorithm nonoptimal actions are played with a relatively low but nonzero probability. The relativeness is determined by the starting vector. When a player becomes in equilibrium he stays so in principle. However, such a partial equilibrium can be disturbed later on.

From the interpretation of the L-H algorithm it is rather straightforward to extend this algorithm in order to find all Nash equilibria in a bi-matrix game with one player, say player 1, having only two actions. Like for the search for positively indexed equilibria we only consider bi-matrix games having an odd number of Nash equilibria. Two arguments are of importance in this respect. First the fact that all Nash equilibria lie in R^2, the best reply set of player 2. The second observation is that in case player 1 has only two actions, R^2 itself is a piecewise linear path. More precisely, it can be parametrized by the probability with which player 1 plays action (1,1). We get $R^2 = \cup_{x_{11}\in[0,1]} R^2(x_{11})$, where

$$R^2(x_{11}) := \{\bar{x} \in R^2 | \bar{x}_{11} = x_{11}\}.$$

Now, for any $x_{11} \in [0,1]$, $R^2(x_{11})$ is either a point or a line segment. In the first case player 2 has one optimal action against $x_1 = (x_{11}, 1-x_{11})$, in the latter case two of his actions are optimal. Remark that we have excluded cases in which more than two actions are optimal for player 2.

In its standard form L-H traces R^2 starting from $x_{11} = 1$ till an NE is found. Now, we extend L-H such that it continues along R^2 after an NE has been found, until x_{11} becomes 0. In this way it finds all Nash equilibria. To implement this consider the algorithm operating on system (6.5.4). Recall that this pivoting algorithm is equivalent to L-H except that it operates on S. An NE is found as soon as x_{11} or μ_{11} becomes zero. In the first case L-H followed the whole R^2 to find the unique NE at which $x_1 = (0,1)$. In the second case the algorithm is continued by letting action (1,2) play the role of (1,1). Note that before μ_{11} became zero, x_{11} was positive whereas (1,2) was optimal for player 1. We continue the algorithm after μ_{11} has become zero by increasing μ_{12} from zero. Then the algorithm generates strategy vectors x at which x_{12} is positive whereas (1,2) is not optimal. The next NE is reached when μ_{12} becomes zero. Then μ_{11} is increased, and so on. We illustrate this again with Example 6.5.1.

Example 6.5.1 (continued). Consider Figure 6.5.1 and system (6.5.4). L-H starts from ((1,0),(0,1,0)) and reaches $x^* = ((3/4,1/4),(0,1/5,4/5))$ in two steps. At x^* for the solution to the other variables in (6.5.4) it holds that $\beta_1 = z_{11}(x^*) = z_{12}(x^*) = 18/5$, $\beta_2 = z_{22}(x^*) = z_{23}(x^*) = 10/4$, and $\mu_{21} = 11/4$. We now continue by increasing μ_{12} from zero. When μ_{12} becomes 1, x_{22} becomes zero. Then μ_{22} is increased till μ_{22} becomes 11/9. At that point μ_{21} becomes zero and the algorithm continues by increasing x_{21} till μ_{12} becomes zero at the NE $\bar{x} = ((4/9,5/9),(1/3,0,2/3))$. We then have that $\mu_{22} = 6/9$. The procedure leaves $\bar{x}$ by increasing μ_{11} from zero. As soon as μ_{11} becomes 2, x_{23} becomes zero. Then μ_{23} is increased till x_{11} becomes zero at the NE $\hat{x} = ((0,1),(1,0,0))$ and the extended L-H algorithm stops.

Thus, in the special case of a (2×m) bi-matrix game we can also find the negatively indexed Nash equilibria in a straightforward manner. For example, the strategy vector $\bar{x}$ in Figure 6.5.1 is positively indexed. Let us view upon the extended version of L-H in another way. Consider again Figure 6.5.1. The best reply set of player 1 consists of three parts; the triangle with vertices C,D, ((1,0),(0,0,1)), the set CDEF, and the polytope with vertices F,E, ((0,1),(0,1,0)), $\hat{x}$. The Nash equilibria are the intersection points of R^1 and R^2. Starting from ((1,0), (0,1,0)) we first intersect R^1 from below at x^*, next from above at $\bar{x}$, and finally at $\hat{x}$. Below CDEF action (1,2) is optimal for player 1 ($\mu_{11} > 0$), whereas $\mu_{12} > 0$ above that plane. Thus, when the algorithm reaches x^* it is quite natural to continue by increasing μ_{12} from zero.

Another method for finding all Nash equilibria in a (2×m) bi-matrix game is a geometric combinatorial method developed by Borm, Gijsberts, and Tijs [1989]. The advantage of their method is that they not only find all Nash equilibria but they can also distinguish Nash equilibria according to their type (perfect, proper and so on). However, although their method gives characterizations of Nash equilibria of certain type it seems rather difficult to implement it for computational purposes, especially when m is large.

Finally, we return to the remark made at the end of Section 6.4 concerning the problem that in general it cannot be assured that our algorithm has found all equilibria. Here, theory developed in Le Van [1982] sheds light. Le Van considers the set of solutions to system (6.5.1) and gives a method to find the exact number of Nash solutions. First, he gives a bounded region in which all solutions lie. Then, that region is subdivided into subregions and the topological degree of a suitable mapping in each subregion is calculated. From that it becomes clear if such a subregion contains a Nash equilibrium. By projecting that region on the strategy space we get information about the location of a Nash equilibrium in the strategy space. Thus, by his method we can determine the approximate location of all equilibria whereas by our method we can find their exact location. However, for practical purposes this combined procedure will be costly because for example the calculation of a topological degree is rather complex.

6.6. Computing Nash equilibria in noncooperative more-person games

In this section we show how to compute Nash equilibria in noncooperative games played by more than two players. More precisely, we consider games with N players, indexed by $j \in \{1,\dots,N\}$, player j having n_j actions. The payoffs to player j are listed in a tensor A^j, $j \in \{1,\dots,N\}$. The payoff to player j in case each player $i \in I_N$ plays action (i,ℓ_i) is denoted by $A^j(\ell_1,\ell_2,\dots,\ell_{j-1},\ell_j,\ell_{j+1},\dots,\ell_N)$. The marginal payoff function z is now a function from S to $\prod_{j=1}^N \mathbb{R}^{n_j}$, with $S = \prod_{j=1}^N S^{n_j-1}$, defined by $z(x) = (z_1(x),\dots,z_N(x))$, with

$$z_{jk}(x) = \Sigma_{\ell_1=1}^{n_1}\dots\Sigma_{\ell_{j-1}=1}^{n_{j-1}}\Sigma_{\ell_{j+1}=1}^{n_{j+1}}\dots\Sigma_{\ell_N=1}^{n_N} A^j(\ell_1,\dots,\ell_{j-1},k,$$

$$\ell_{j+1},\dots,\ell_N)\prod_{i\neq j} x_{i\ell_i}, \quad (j,k) \in I(j),\ j \in I_N.$$

Similar to a bi-matrix game (cf. (6.1.2)), a strategy vector x^* in S is defined to be an NE if

$$x_j \cdot z_j(x^*) \leqq x_j^* \cdot z_j(x^*) \quad , \ x_j \in S^{n_j-1} \ , \ j \in I_N. \tag{6.6.1}$$

The most well-known methods for finding an NE in an N-person game are the simplicial algorithms. For an exposition and some computational results we refer to Doup and Talman [1987]. At the end of this section we compare the results obtained with the simplicial methods and the results we obtained with the procedure discussed here. Concerning other methods we refer to the articles of Rosenmüller [1971] and Wilson [1971]. They both pretend to present an algorithm. However, both articles are theoretical of nature and of hardly any use for practical implementation. Rosenmüller argues that each nondegenerated N-person game has an odd number of isolated Nash equilibria. The reasoning is similar to that in Lemke and Howson [1964] for bi-matrix games; there is a path connecting the constructed starting vector and an NE whereas the other Nash equilibria are pairwise connected by paths. However, he gives no procedure how to follow these paths. The latter is crucial because these paths are in general not linear. Wilson [1971] proves the existence of a path leading to an NE in an N-person game. This path finds in succession an equilibrium for each of certain related k-person games for k increasing from 1 to N. An equilibrium for the k-person game is then the starting point of a path leading to an NE of a (k+1)-person game and so on. However, also here merely the existence of the path is proved and no method is given to follow it.

The method we present here is a generalization of the algorithm for solving bi-matrix games that has been discussed in Section 6.2. Again we can rewrite (6.6.1) to obtain that x^* is an NE iff

$$z_{jk}(x^*) = \max_h z_{jh}(x^*) \text{ when } x^*_{jk} > 0, \ (j,k) \in I. \tag{6.6.2}$$

Expression (6.6.2) is equivalent to (6.1.3). However, contrary to the function z for a bi-matrix game, the marginal payoff function is not linear in the more general case. In fact z is multilinear. It appears that we can use the algorithm of Section 6.2 in an indirect manner. The general idea is the following. First, we choose an arbitrary starting strategy vector v in S. Next, we linearize the marginal payoff function z around v by a first-order Taylor expansion. We then extend that function on S and apply a slightly more general version of the algorithm of Section 6.2 to find a stationary point of that function. We then repeat the procedure starting from the vector obtained in this way. Thus, we now linearize z around that vector, extend the linear function to S, and again apply our procedure. We continue till a certain accuracy of approximation is obtained. Two remarks are of importance here. First, we cannot guarantee that this procedure converges towards a Nash equilibrium. Although it holds that the algorithm of Section 6.2 converges for each linearized version of z, the global convergence of the whole procedure is questionable. The second point is that we can only <u>approximate</u> an NE here, whereas we obtained an <u>exact</u> solution in case of a bi-matrix game. Again, in each round we obtain an exact stationary point for the linearized marginal payoff function. But this is only an approximate stationary point for the marginal payoff function itself. In this way we approximate an NE by solving a sequence of linear stationary point problems. Therefore we sometimes refer to our method as the SLSPP-method (sequence of linear stationary point problems). This procedure has been motivated by Mathiessen [1985a] who solves an NLCP by a sequence of LCP's. In practice this method works quite satisfactory, although global convergence is not guaranteed. Contrary to our method, his procedure may even fail to solve a specific LCP (see Mathiessen [1985b]).

We remark that in some sense the method sketched above works similar as a simplicial algorithm. Such a method finds in each round an exact solution for a piecewise linear approximation of the original function. Thus, then the function is linearized on each simplex of some simplicial subdivision, whereas here we take a linear approximation on the whole set. The advantage of simplicial algorithms is that their global convergence is guaranteed. However, especially in the later rounds simplicial algorithms may need more pivot steps to reach a solution. In practice it seems best to combine both methods by applying a simplicial algorithm during the first rounds, and later on the method discussed here. We observe that our method can be used for solving an arbitrary stationary point problem on the simplotope.

We now review our procedure in more detail. Let us denote the total number of actions by n, i.e., $n = \Sigma_{j=1}^{N} n_j$. In the sequel we view strategy vectors as vectors in $\mathbb{R}^n$. First we choose an arbitrary starting vector v in the interior of the strategy space. Then we linearize z around v to obtain the function $z^v : S \to \Pi_{j=1}^{N} \mathbb{R}^{n_j}$ defined by

$$z^v(x) = z(v) + Dz(v)(x - v), \tag{6.6.3}$$

where Dz(v) is the (n×n)-matrix of derivatives of z at v. More precisely, the (p,q)-th element of Dz(v) equals $\delta z_{jk}(v)/\delta x_{ih}$, where $p = \Sigma_{\ell=1}^{j-1} n_\ell + k$ and $q = \Sigma_{\ell=1}^{i-1} n_\ell + h$. Observe that this element is zero if $(i,h) \in I(j)$. Now we apply the algorithm of Section 6.2 to z^v. Of course, we need to generalize that algorithm for application to problems on a simplotope instead of the product of two simplices. Analogous to Section 6.2 we follow a path of strategy vectors x in S such that for some $T \subset I$

$$x = bv + \Sigma_{(j,k)\in T}\lambda_{jk}e(j,k),$$

and

$$z^v_{jk}(x) = \max_{(j,\ell)\in I(j)} z^v_{j\ell}(x) = \beta_j \text{ for } (j,k) \in T, \tag{6.6.4}$$

with $\Sigma_{(j,k)\in T_j}\lambda_{jk} = 1\text{-}b$ for $j \in I_N$, $\lambda_{jk} \geq 0$ for $(j,k) \in T$, and $b \in [0,1]$.

Combining (6.6.3) and (6.6.4) gives the following linear system of equations

$$bDz(v)v + \Sigma_{(j,k)\in T}\lambda_{jk}D^{jk}z(v) + \Sigma_{(i,h)\notin T}\mu_{ih}e(i,h) - \Sigma_{j\in I_N}\beta_j\bar{e}_j = Dz(v)v - z(v)$$

$$b + \Sigma_{(j,k)\in T_j}\lambda_{jk} = 1 \ , \ j \in I_N, \tag{6.6.5}$$

where $D^{jk}z(v)$ denotes the $(\Sigma_{\ell=1}^{j-1}n_\ell + k)$-th column of $Dz(v)$ and $\bar{e}_j$ denotes a vector in $\mathbb{R}^n$ with ones on the places $\Sigma_{\ell=1}^{j-1}n_\ell + k$, $k \in \{1,\dots,n_j\}$, and zero elsewhere. Note, that this system is analogous to (6.2.5). Similar to Section 6.2 we obtain the final system by substituting $b' = 1\text{-}b$ and for $j \in I_N$, λ_{jk_j} by $b' - \Sigma_{k\neq k_j}\lambda_{jk}$ where k_j is an index for which $z_{jk_j}(v) = \max_\ell z_{j\ell}(v)$. In this way, we obtain a system of n equations with n+1 variables. More precise, we get (cf. (6.2.7))

$$b'(\Sigma_{j\in I_N}D^{jk_j}z(v) - Dz(v)v) + \Sigma_{\substack{(j,k)\in T \\ k\neq k_j}}\lambda_{jk}(D^{jk}z(v) - D^{jk_j}z(v))$$

$$+ \Sigma_{(i,h)\notin T}\mu_{ih}e(i,h) - \Sigma_{j\in I_N}\beta_j\bar{e}_j = -z(v), \qquad (6.6.6)$$

with additional restrictions $\lambda_{jk} \geqq 0$ for $(j,k) \in T_j\backslash\{(j,k_j)\}$, $\Sigma_{k\neq k_j}\lambda_{jk} \leqq b'$ for $j \in I_N$, $0 \leqq b' \leqq 1$, and $\mu_{ih} \geqq 0$ for $(i,h) \notin T$. Observe that (6.6.6) is equivalent to (6.2.7) in case N = 2 because in that case

$$Dz(v) = \begin{bmatrix} 0 & A \\ B^T & 0 \end{bmatrix}, \quad v \in S.$$

The working of the algorithm for the N-person game is essentially identical to that of Section 6.2. For convenience we list below without further comments the nondegeneracy assumption needed and the steps of the algorithm. Here by λ and μ we denote the n-vectors $(\lambda_1,\ldots,\lambda_N)$ and $(\mu_1,\ldots,\mu_N)$, respectively, and by β we denote the N-vector $(\beta_1,\ldots,\beta_N)$. The other notation used is similar to that one in Section 6.2.

Assumption 6.6.1 (nondegeneracy assumption). At each solution (b',λ,μ,β) of (6.6.6) at most one of the constraints $0 \leqq b' \leqq 1$, $\lambda_{jk} \geqq 0$ for $(j,k) \in T^1$, $b' \geqq \Sigma_{(j,k)\in T_j^1}\lambda_{jk}$, $\mu_{ih} \geqq 0$ for $(i,h) \notin T$, is binding, unless $b' = 1$ or $v_{ih} = 0$ for all $(i,h) \notin T$.

Step 0 [Initialization].

Choose an arbitrary vector v in S. If v is a Nash equilibrium then the algorithm stops. Else choose a measure of inaccuracy δ, a maximum number of rounds $\bar{t}$, and set t equal to 1.

Step 1.

If $t > \bar{t}$ then the algorithm stops. No approximate Nash equilibrium has been found within $\bar{t}$ rounds. Else construct the first-order Taylor expansion z^v of z at v as in (6.6.3). Calculate for $j \in I_N$ the unique index (j,k_j) for which $z_{jk_j}(v) = \max_h z_{jh}(v)$. Furthermore, set $T^1 = \emptyset$, $b' = 0$, $\beta_j = z_{jk_j}(v)$ for $j \in I_N$, $\mu_{ih} = \beta_i - z_{ih}(v)$ for $h \neq k_i$ and $i \in I_N$. Increase b' from 0 in (6.6.6) and go to Step 2.

Step 2.

a. If b' becomes 1 then let the solution of (6.6.6) be $(1,\lambda^*,\mu^*,\beta^*)$. The vector x^*, with $x^* = \Sigma_{(j,k)\in T}\lambda^*_{jk}e(j,k)$ is a solution to the SPP of z^v on S. If $\max_{(j,k)}(z_{jk}(x^*)-x^*_j \cdot z_j(x^*)) < \delta$ then x^* is an approximate Nash equilibrium and the algorithm stops after round t. Else t becomes t+1 and return to Step 1 with v equal to x^*.

b. If λ_{jk} becomes 0 for some $(j,k) \in T^1$ then T^1 becomes $T^1\backslash\{(j,k)\}$ and go to Step 3a.

c. If $\Sigma_{(j,k)\in T^1_j} \lambda_{jk}$ becomes equal to b' for some $j \in I_N$ then λ_{jk_j} becomes 0. Go to Step 3b.

d. If μ_{ih} becomes zero for some $(i,h) \notin T$ then go to Step 4.

Step 3.

a. Increase the complementary variable μ_{jk} from zero by pivoting into system (6.6.6) the column $e(j,k)$. Return to Step 2.

b. Substitute the largest λ_{jk}, say $\lambda_{j\ell}$, by $b'-\Sigma_{(j,h)\in T_j, h\neq k_j,\ell}\lambda_{jh}$. Increase μ_{jk_j} from zero by pivoting the column $e(j,k_j)$ into system (6.6.6). T^1 becomes $T^1\backslash\{(j,\ell)\}$, k_j becomes ℓ, and return to Step 2.

Step 4.

a. If additionally $v_{jk} = 0$ for all $(j,k) \notin T \cup \{(i,h)\}$ then let the solution be $(b'^*,\lambda^*,\mu^*,\beta^*)$. The vector $x^* = (1-b'^*)v + \Sigma_{(j,k)\in T} \lambda^*_{jk}e(j,k)$ is a solution to the SPP of z^v on S. If $\max_{(j,k)}(z_{jk}(x^*)-x^*_j \cdot z_j(x^*)) < \delta$ then x^* is an approximate Nash equilibrium and the algorithm stops after round t, else t becomes t+1, and return to Step 1 with v equal to x^*.

b. Otherwise increase the complementary variable λ_{ih} from zero by pivoting its related column into system (6.6.6). T^1 becomes $T^1 \cup \{(i,h)\}$ and return to Step 2.

We illustrate the working of the procedure by an example.

Example 6.6.1. We consider a noncooperative 3-person game, in which each player has two actions. The payoffs are listed below.

(1,1)

	(2,1)	(2,2)
(3,1)	(1,3,-2)	(2,1,4)
(3,2)	(-1,-2,3)	(4,5,-6)

(1,2)

	(2,1)	(2,2)
(3,1)	(1,-3,2)	(1,1,-3)
(3,2)	(1,-3,4)	(-5,1,-2)

The left matrix entails the payoffs in case player 1 plays his first action, the right matrix corresponds to strategy vectors at which player 1 plays his second action. Each entry in the matrix consists of three elements corresponding to the payoffs for each player. For example, if player 1 plays (1,2), player 2 plays (2,2) and 3 plays (3,1) then player 1 and 2 get a payoff of 1 whereas player 3 gets -3.

Let us apply our algorithm starting from $v = ((1/2,1/2),(1/2,1/2),(1/2,1/2))$. Linearizing the marginal payoff function z around v gives the function z^v defined by $z^v(x) = z(v) + Dz(v).(x - v)$

$$= \begin{bmatrix} 1.5 \\ -0.5 \\ -1.25 \\ 2 \\ 0.25 \\ -0.25 \end{bmatrix} + \begin{bmatrix} 0 & 0 & 0 & 3 & 1.5 & 1.5 \\ 0 & 0 & 1 & -2 & 1 & -2 \\ 0.5 & -3 & 0 & 0 & 0 & -2.5 \\ 3 & 1 & 0 & 0 & 1 & 3 \\ 1 & -0.5 & 0 & 0.5 & 0 & 0 \\ -1.5 & 1 & 3.5 & -4 & 0 & 0 \end{bmatrix} \begin{bmatrix} x_{11}-0.5 \\ x_{12}-0.5 \\ x_{21}-0.5 \\ x_{22}-0.5 \\ x_{31}-0.5 \\ x_{32}-0.5 \end{bmatrix}.$$

At $x = v$, $z^v(x)$ equals $z(v)$ and the optimal action for player 1 is (1,1), for player 2 action (2,2), for player 3 action (3,1). Thus, from the start the related probabilities are increased by increasing b' from zero (Step 1). After one step a stationary point for z^v is found because b' becomes 1 (Step 2a). It is easily checked that the vector $\bar{x} = ((1,0),(0,1),(1,0))$ is indeed an NE related to the marginal payoff function z^v. However $z(\bar{x})$ equals ((2,1),(3,1),(4,-6)), i.e. $\bar{x}$ is not an NE for the original game because player 2 is not in equilibrium at $\bar{x}$. In the second round we linearize z around $\bar{x}$ and obtain with $v = \bar{x}$ that

$$z^v(x) = z(v) + Dz(v).(x-v) = \begin{bmatrix} 2 \\ 1 \\ 3 \\ 1 \\ 4 \\ -6 \end{bmatrix} + \begin{bmatrix} 0 & 0 & 1 & 2 & 2 & 4 \\ 0 & 0 & 1 & 1 & 1 & -5 \\ 3 & -3 & 0 & 0 & 3 & -2 \\ 1 & 1 & 0 & 0 & 1 & 5 \\ 4 & -3 & -2 & 4 & 0 & 0 \\ -6 & -2 & 3 & -6 & 0 & 0 \end{bmatrix} \begin{bmatrix} x_{11}-1 \\ x_{12} \\ x_{21} \\ x_{22}-1 \\ x_{31}-1 \\ x_{32} \end{bmatrix}.$$

At the start we have that $\beta_1 = z_{11}(v) = 2$, $\beta_2 = z_{21}(v) = 3$, and $\beta_3 = z_{31}(v) = 4$. Furthermore $\mu_{12} = 1$, $\mu_{22} = 2$, and $\mu_{32} = 10$. Then b' is increased from 0 (Step 1). If b' becomes 2/3 then μ_{32} becomes 0 (Step 2d). For the solution to (6.6.6) it further holds that $\beta = (\beta_1, \beta_2, \beta_3) = (4/3, 3, 0)$, $\mu_{12} = 1/3$, $\mu_{22} = 2$. Thus, now λ_{32} is increased (Step 4) till 2/9 when μ_{22} becomes zero (Step 2d). For the solution to (6.6.6) we have $\beta = (16/9, 17/9, 0)$, $\mu_{12} = 19/9$, $\lambda_{31} = 4/9$, b' = 2/3 with corresponding vector $x^* = ((1,0), (2/3,1/3), (7/9,2/9))$. Since $v_{jk} = 0$ for all (j,k) such that $\mu_{jk} > 0$, x^* is an equilibrium for z^v (Step 4a). It is easily verified that x^* is also an NE for the original game.

Finally we compare the speed of our algorithm with simplicial algorithms. We applied our algorithm to the three games given in Doup and Talman [1987] and compared the results with those given in Doup [1988, ch. 11]. These three different simplicial algorithms are applied to the games. In Table 6.6.1 we compare our results with the best results obtained by the simplicial algorithms. It turns out that our algorithm is at least as good.

Game	Simplicial algorithms		SLSPP algorithm	
	LP	ν	LP	ν
1	35	4	28	4
2	14	1	6	2
3	14	3	14	3

Table 6.6.1. Computational results obtained for noncooperative more-person games. LP denotes the number of linear programming pivoting steps whereas ν denotes the number of rounds.

References

Bastian, M.L. [1976], "Another note on bimatrix games", Mathematical Programming 11, 299-300.

Bernheim, D. [1984], "Rationalizable strategic behaviour", Econometrica 52, 1007-1028.

Borm, P., A. Gijsberts, and S.H. Tijs [1989], "A geometric-combinatorial approach to bimatrix games", Methods of Operations Research 59, 199-209.

Brouwer, L.E.J. [1912], "Über Abbildung von Mannigfaltigkeiten", Mathematische Annalen 71, 97-115.

Damme, E.E.C. van [1983], Refinements of the Nash Equilibrium Concept, Lecture Notes in Economics and Mathematical Systems 219, Springer-Verlag, Berlin.

Dang, C. [1991], The D_1-Triangulation in Simplicial Algorithms, Ph.D. Thesis, Tilburg University, Tilburg, The Netherlands.

Debreu, G. [1974], "Excess demand functions", Journal of Mathematical Economics 1, 15-23.

Doup, T.M. [1988], Simplicial Algorithms on the Simplotope, Lecture Notes in Economics and Mathematical Systems 318, Springer-Verlag, Berlin.

Doup, T.M., A.H. van den Elzen, and A.J.J. Talman [1987], "Simplicial algorithms for solving the nonlinear complementarity problem on the simplotope", in: A.J.J. Talman and G. van der Laan (eds.), The Computation and Modelling of Economic Equilibria, North-Holland, Amsterdam, pp. 125-154.

Doup, T.M., A.H. van den Elzen, and A.J.J. Talman [1989], "Homotopy interpretation of price adjustment processes", Research Memorandum 384, Tilburg University, Tilburg, The Netherlands.

Doup, T.M., G. van der Laan, and A.J.J. Talman [1987], "The $(2^{n+1}-2)$-ray algorithm: a new simplicial algorithm to compute economic equilibria", Mathematical Programming 39, 241-252.

Doup, T.M. and A.J.J. Talman [1987], "A new variable dimension simplicial algorithm to find equilibria on the product space of unit simplices", Mathematical Programming 37, 319-355.

Eaves, B.C. [1987], "Thoughts on computing market equilibrium with SLCP", in: A.J.J. Talman and G. van der Laan (eds.), The Computation and Modelling of Economic Equilibria, North-Holland, Amsterdam, pp. 1-17.

Elzen, A.H. van den [1990], "Interpretation and generalization of the Lemke-Howson algorithm", Methods of Operations Research 60, 337-345.

Elzen, A.H. van den, and G. van der Laan [1986], "A price adjustment process for an economy with a block-diagonal pattern", Research Report 166, Free University, Amsterdam, The Netherlands.

Elzen, A.H. van den, and G. van der Laan [1989], "Price adjustment in a two-country model", Research Memorandum 391, Tilburg University, Tilburg, The Netherlands.

Elzen, A.H. van den, G. van der Laan, and A.J.J. Talman [1985], "Adjustment processes for finding equilibria on the simplotope", Research Memorandum 196, Tilburg University, Tilburg, The Netherlands.

Elzen, A.H. van den, G. van der Laan, and A.J.J. Talman [1990], "An adjustment process for an exchange economy with linear production technologies", CentER Discussion Paper 9015, Tilburg University, Tilburg, The Netherlands.

Elzen, A.H. van den, and A.J.J. Talman [1986], "A new strategy-adjustment process for computing a Nash equilibrium in a noncooperative more-person game", Methods of Operations Research 54, pp. 469-481.

Elzen, A.H. van den, and A.J.J. Talman [1991], "A procedure for finding Nash equilibria in bi-matrix games", ZOR-Methods and Models of Operations Research 35, 27-43.

Elzen, A.H. van den, and A.J.J. Talman [1992], "Finding a Nash equilibrium in noncooperative N-person games by solving a sequence of linear stationary point problems", Research Memorandum, Tilburg University, Tilburg, The Netherlands.

Gale, D. [1986], "Bargaining and competition part I: characterization", Econometrica 54, 785-806.

Garcia, C.B. and W.I. Zangwill [1981], Pathways to Solutions, Fixed Points and Equilibria, Prentice-Hall, Englewood Cliffs, N.J., U.S.A.

Geldrop, J.H. van [1981], A Mathematical Theory of Pure Exchange Economies Without the No-Critical-Point Hypothesis, Mathematical Centre Tracts 140, Mathematisch Centrum, Amsterdam, The Netherlands.

Guillemin, V. and A. Pollack [1974], Differential Topology, Prentice-Hall, Englewood Cliffs, N.J., U.S.A.

Hahn, F. [1982], "Stability", in: K.J. Arrow and M.D. Intriligator (eds.), Handbook of Mathematical Economics II, North-Holland, Amsterdam, pp. 745-793.

Hurwicz, L. [1960], "Optimality and informational efficiency in resource allocation processes", in: K.J. Arrow, S. Karlin, and P. Suppes (eds.), Mathematical Methods in Social Sciences, Stanford University Press, Stanford, U.S.A.

Hurwicz, L. [1986], "On informational decentralization and efficiency in resource allocation mechanisms", in: S. Reiter (ed.), MAA Studies in Mathematics 25, Studies in Mathematical Economics, The Mathematical Association of America, pp. 238-250.

Hurwicz, L., R. Radner, and S. Reiter [1975], "A stochastic decentralized resource allocation mechanism, part I", Econometrica 43, 187-223.

Kamiya, K. [1984], "On the existence and uniqueness of general equilibrium prices", International Economic Review 25, 79-83.

Kamiya, K. [1990], "A globally stable price adjustment process", Econometrica 58, 1481-1485.

Keenan, D. [1981], "Further remarks on the global Newton method", Journal of Mathematical Economics 8, 159-166.

Keenan, D. and T. Rader [1985], "Market dynamics and the law of demand", Econometrica 53, 465-471.

Kehoe, T.J.. [1985], "Multiplicity of equilibria and comparative statics", The Quarterly Journal of Economics 100, 119-147.

Kojima, M. and Y. Yamamoto [1982], "Variable dimension algorithms: basic theory, interpretations and extensions of some existing methods", Mathematical Programming 24, 177-215.

Kremers, H. [1991], Pivoting Algorithms for Complementarity Problems in Economics, Ph.D.-Thesis, Tilburg University, Tilburg, The Netherlands.

Kuhn, H.W. [1953], "Extensive games and the problem of information", Annals of Mathematics Studies 28, 193-216.

Kuhn, H.W. [1961], "An algorithm for equilibrium points in bimatrix games", Proceedings of the National Academy of Sciences 47, 1656-1662.

Laan, G. van der [1980], Simplicial Fixed Point Algorithms, Mathematical Centre Tracts 129, Mathematisch Centrum, Amsterdam, The Netherlands.

Laan, G. van der [1985], "The computation of general equilibrium in economies with a block diagonal pattern", Econometrica 53, 659-665.

Laan, G. van der [1988], "Het spel van algemeen evenwicht", Vrije Universiteit Amsterdam, The Netherlands.

Laan, G. van der, and A.J.J. Talman [1979], "A restart algorithm for computing fixed points without an extra dimension", Mathematical Programming 17, 74-84.

Laan, G. van der, and A.J.J. Talman [1983], "Interpretation of a variable dimension fixed point algorithm with an artificial level", Mathematics of Operations Research 8, 86-99.

Laan, G. van der, and A.J.J. Talman [1987a], "Adjustment processes for finding economic equilibria", in: A.J.J. Talman and G. van der Laan (eds.), The Computation and Modelling of Economic Equilibria, North-Holland, Amsterdam, pp. 85-124.

Laan, G. van der, and A.J.J. Talman [1987b], "Computing economic equilibria by variable dimension algorithms: state of the art", Research Memorandum 270, Tilburg University, Tilburg, The Netherlands.

Lemke, C.E. and J.T. Howson [1964], "Equilibrium points of bi-matrix games", SIAM Journal of Applied Mathematics 12, 413-423.

Le Van, C. [1982], "Topological degree and number of Nash equilibrium points of bimatrix games", Journal of Optimization Theory and Applications 37, 355-369.

Mangasarian, O.L. [1964], "Equilibrium points of bimatrix games", SIAM Journal of Applied Mathematics 12, 778-780.

Mansur, A. and J. Whalley [1982], "A decomposition algorithm for general equilibrium computation with application to international trade models", Econometrica 50, 1547-1557.

Mathiesen, L. [1985a], "Computation of economic equilibria by a sequence of linear complementarity problems", Mathematical Programming Study 23, 144-162.

Mathiesen, L. [1985b], "Computational experience in solving equilibrium models by a sequence of linear complementarity problems", Operations Research 33, 1225-1250.

Mathiesen, L. [1987], "An algorithm based on a sequence of linear complementarity problems applied to a walrasian equilibrium model: an example", Mathematical Programming 37, 1-18.

Milnor, J. [1965], Topology from a Differential Viewpoint, University of Virginia Press, U.S.A.

Pearce, D. [1984], "Rationalizable strategic behaviour and the problem of perfection", Econometrica 52, 1029-1040.

Rosenmüller, J. [1971], "On a generalization of the Lemke-Howson algorithm to noncooperative N-person games", SIAM Journal of Applied Mathematics 21, 73-79.

Rubenstein, A. [1982], "Perfect equilibrium in a bargaining model", Econometrica 50, 97-109.

Saari, D.G. [1985], "Iterative price mechanisms", Econometrica 53, 1117-1131.

Saari, D.G. and C.P. Simon [1978], "Effective price mechanisms", Econometrica 46, 1097-1125.

Samuelson, P.A., [1947], Foundations of Economic Analysis, Harvard University Press, Cambridge, U.S.A.

Scarf, H. [1960], "Some examples of global instability of the competitive equilibrium", International Economic Review 1, 157-172.

Scarf, H. [1973], The Computation of Economic Equilibria, Yale University Press, U.S.A.

Selten, R. [1975], "Reexamination of the perfectness concept for equilibrium points in extensive games", International Journal of Game Theory 4, 25-55.

Shapley, L.S. [1974], "A note on the Lemke-Howson algorithm", Mathematical Programming Study 1, 175-189.

Shubik, M. [1984], A Game Theoretic Approach to Political Economy, MIT Press, Cambridge, U.S.A.

Smale, S. [1976], "A convergent process of price adjustment and global Newton methods", Journal of Mathematical Economics 3, 107-120.

Talman, A.J.J. [1980], Variable Dimension Fixed Point Algorithms and Triangulations, Mathematical Centre Tracts 128, Mathematisch Centrum, Amsterdam, The Netherlands.

Talman, A.J.J. and Y. Yamamoto [1989], "A simplicial algorithm for stationary point problems on polytopes", Mathematics of Operations Research 14, 383-399.

Todd, M.J. [1976], The Computation of Fixed Points and Applications, Lecture Notes in Economics and Mathematical Systems 124, Springer-Verlag, Berlin.

Todd, M.J. [1978], "Bimatrix games - an addendum", Mathematical Programming 14, 112-115.

Todd, M.J. [1979], "A note on computing equilibria in economies with activity analysis models of production", Journal of Mathematical Economics 6, 135-144.

Vorob'ev, N.N. [1958], "Equilibrium points in bimatrix games" (translated by R. De Marr), Theory of Probability and its Applications, III, 297-309.

Uzawa, H. [1962], "On the stability of Edgeworth's barter process", International Economic Review 3, 218-232.

Walras, L. [1874], Éléments d'Economie Politique Pure, L. Corbaz and Company, Lausanne, Switzerland.

Wilson, R. [1971], "Computing equilibria of N-person games", SIAM Journal of Applied Mathematics 21, 80-87.

Winkels, H.-M. [1979], "An algorithm to determine all equilibrium points of a bimatrix game", in: O. Moeschlin and D. Pallaschke (eds.), Game Theory and Related Topics, North-Holland, Amsterdam, pp. 1307-1322.

Springer-Verlag and the Environment

We at Springer-Verlag firmly believe that an international science publisher has a special obligation to the environment, and our corporate policies consistently reflect this conviction.

We also expect our business partners – paper mills, printers, packaging manufacturers, etc. – to commit themselves to using environmentally friendly materials and production processes.

The paper in this book is made from low- or no-chlorine pulp and is acid free, in conformance with international standards for paper permanency.

Lecture Notes in Economics and Mathematical Systems

For information about Vols. 1–223 please contact your bookseller or Springer-Verlag

Vol. 224: M. Kiy, Ein disaggregiertes Prognosesystem fur die Bundesrepublik Deutschland. XVIII, 276 Seiten. 1984.

Vol. 225: T.R. von Ungern-Sternberg, Zur Analyse von Märkten mit unvollständiger Nachfragerinformaton. IX, 125 Seiten. 1984.

Vol. 226: Selected Topics in Operations Research and Mathematical Economics. Proceedings, 1963. Edited by G. Hammer and D. Pallaschke IX, 478 pages. 1984.

Vol. 227: Risk and Capital. Proceedings, 1983. Edited by G. Bamberg and K. Spremann VII, 306 pages. 1984.

Vol. 228: Nonlinear Models of Fluctuating Growth. Proceedings, 1983. Edited by R.M. Goodwin, M. Krüger and A. Vercelli. XVII, 277 pages. 1984.

Vol. 229: Interactive Decision Analysis. Proceedings, 1983. Edited by M. Grauer and A.P. Wierzbicki. VIII, 269 pages. 1984.

Vol. 230: Macro-Economic Planning with Conflicting Goals. Proceedings, 1982. Edited by M. Despontin, P. Nijkamp and J. Spronk. VI, 297 pages. 1984.

Vol. 231: G.F. Newell, The M/M/8 Service System with Ranked Servers in Heavy Traffic. XI, 126 pages. 1984.

Vol. 232: L. Bauwens, Bayesian Full Information Analysis of Simultaneous Equation Models Using Integration by Monte Carlo. VI, 114 pages. 1984.

Vol. 233: G. Wagenhals, The World Copper Market. XI, 190 pages. 1984.

Vol. 234: B.C. Eaves, A Course in Triangulations for Solving Equations with Deformations. III, 302 pages. 1984.

Vol. 235: Stochastic Models in Reliability Theory Proceedings, 1984. Edited by S. Osaki and Y. Hatoyama. VII, 212 pages. 1984.

Vol. 236: G. Gandolfo, P.C. Padoan, A Disequilibrium Model of Real and Financial Accumulation in an Open Economy. VI, 172 pages. 1984.

Vol. 237: Misspecification Analysis. Proceedings, 1983. Edited by T.K. Dijkstra. V, 129 pages. 1984.

Vol. 238: W. Domschke, A. Drexl, Location and Layout Planning. IV, 134 pages. 1985.

Vol. 239: Microeconomic Models of Housing Markets. Edited by K. Stahl. VII, 197 pages. 1985.

Vol. 240: Contributions to Operations Research. Proceedings, 1984. Edited by K. Neumann and D. Pallaschke. V, 190 pages. 1985.

Vol. 241: U. Wittmann, Das Konzept rationaler Preiserwartungen. XI, 310 Seiten. 1985.

Vol. 242: Decision Making with Multiple Objectives. Proceedings, 1984. Edited by Y.Y. Haimes and V. Chankong. XI, 571 pages. 1985.

Vol. 243: Integer Programming and Related Areas. A Classified Bibliography 1981–1984. Edited by R. von Randow. XX, 386 pages. 1985.

Vol. 244: Advances in Equilibrium Theory. Proceedings, 1984. Edited by C.D. Aliprantis, O. Burkinshaw and N.J. Rothman. II, 235 pages. 1985.

Vol. 245: J.E.M. Wilhelm, Arbitrage Theory. VII, 114 pages. 1985.

Vol. 246: P.W. Otter, Dynamic Feature Space Modelling, Filtering and Self-Tuning Control of Stochastic Systems. XIV, 177 pages.1985.

Vol. 247: Optimization and Discrete Choice in Urban Systems. Proceedings, 1983. Edited by B.G. Hutchinson, P. Nijkamp and M. Batty VI, 371 pages. 1985.

Vol. 248: Pural Rationality and Interactive Decision Processes. Proceedings, 1984. Edited by M. Grauer, M. Thompson and A.P. Wierzbicki. VI, 354 pages. 1985.

Vol. 249: Spatial Price Equilibrium: Advances in Theory, Computation and Application. Proceedings, 1984. Edited by P.T. Harker. VII, 277 pages. 1985.

Vol. 250: M. Roubens, Ph. Vincke, Preference Modelling. VIII, 94 pages. 1985.

Vol. 251: Input-Output Modeling. Proceedings, 1984. Edited by A. Smyshlyaev. VI, 261 pages. 1985.

Vol. 252: A. Birolini, On the Use of Stochastic Processes in Modeling Reliability Problems. VI, 105 pages. 1985.

Vol. 253: C. Withagen, Economic Theory and International Trade in Natural Exhaustible Resources. VI, 172 pages. 1985.

Vol. 254: S. Müller, Arbitrage Pricing of Contingent Claims. VIII, 151 pages. 1985.

Vol. 255: Nondifferentiable Optimization: Motivations and Applications. Proceedings, 1984. Edited by V.F. Demyanov and D. Pallaschke. VI, 350 pages. 1985.

Vol. 256: Convexity and Duality in Optimization. Proceedings, 1984. Edited by J. Ponstein. V, 142 pages. 1985.

Vol. 257: Dynamics of Macrosystems. Proceedings, 1984. Edited by J.-P. Aubin, D. Saari and K. Sigmund. VI, 280 pages. 1985.

Vol. 258: H. Funke, Eine allgemeine Theorie der Polypol- und Oligopolpreisbildung. III, 237 pages. 1985.

Vol. 259: Infinite Programming. Proceedings, 1984. Edited by E.J. Anderson and A.B. Philpott. XIV, 244 pages. 1985.

Vol. 260: H.-J. Kruse, Degeneracy Graphs and the Neighbourhood Problem. VIII, 128 pages. 1986.

Vol. 261: Th.R. Gulledge, Jr., N.K. Womer, The Economics of Made-to-Order Production. VI, 134 pages. 1986.

Vol. 262: H.U. Buhl, A Neo-Classical Theory of Distribution and Wealth. V, 146 pages. 1986.

Vol. 263: M. Schäfer, Resource Extraction and Market Struucture. XI, 154 pages. 1986.

Vol. 264: Models of Economic Dynamics. Proceedings, 1983. Edited by H.F. Sonnenschein. VII, 212 pages. 1986.

Vol. 265: Dynamic Games and Applications in Economics. Edited by T. Basar. IX, 288 pages. 1986.

Vol. 266: Multi-Stage Production Planning and Inventory Control. Edited by S. Axsäter, Ch. Schneeweiss and E. Silver. V, 264 pages. 1986.

Vol. 267: R. Bemelmans, The Capacity Aspect of Inventories. IX, 165 pages. 1986.

Vol. 268: V. Firchau, Information Evaluation in Capital Markets. VII, 103 pages. 1986.

Vol. 269: A. Borglin, H. Keiding, Optimality in Infinite Horizon Economies. VI, 180 pages. 1986.

Vol. 270: Technological Change, Employment and Spatial Dynamics. Proceedings, 1985. Edited by P. Nijkamp. VII, 466 pages. 1986.

Vol. 271: C. Hildreth, The Cowles Commission in Chicago, 1939–1955. V, 176 pages. 1986.

Vol. 272: G. Clemenz, Credit Markets with Asymmetric Information. VIII, 212 pages. 1986.

Vol. 273: Large-Scale Modelling and Interactive Decision Analysis. Proceedings, 1985. Edited by G. Fandel, M. Grauer, A. Kurzhanski and A.P. Wierzbicki. VII, 363 pages. 1986.

Vol. 274: W.K. Klein Haneveld, Duality in Stochastic Linear and Dynamic Programming. VII, 295 pages. 1986.

Vol. 275: Competition, Instability, and Nonlinear Cycles. Proceedings, 1985. Edited by W. Semmler. XII, 340 pages. 1986.

Vol. 276: M.R. Baye, D.A. Black, Consumer Behavior, Cost of Living Measures, and the Income Tax. VII, 119 pages. 1986.

Vol. 277: Studies in Austrian Capital Theory, Investment and Time. Edited by M. Faber. VI, 317 pages. 1986.

Vol. 278: W.E. Diewert, The Measurement of the Economic Benefits of Infrastructure Services. V, 202 pages. 1986.

Vol. 279: H.-J. Büttler, G. Frei and B. Schips, Estimation of Disequilibrium Modes. VI, 114 pages. 1986.

Vol. 280: H.T. Lau, Combinatorial Heuristic Algorithms with FORTRAN. VII, 126 pages. 1986.

Vol. 281: Ch.-L. Hwang, M.-J. Lin, Group Decision Making under Multiple Criteria. XI, 400 pages. 1987.

Vol. 282: K. Schittkowski, More Test Examples for Nonlinear Programming Codes. V, 261 pages. 1987.

Vol. 283: G. Gabisch, H.-W. Lorenz, Business Cycle Theory. VII, 229 pages. 1987.

Vol. 284: H. Lütkepohl, Forecasting Aggregated Vector ARMA Processes. X, 323 pages. 1987.

Vol. 285: Toward Interactive and Intelligent Decision Support Systems. Volume 1. Proceedings, 1986. Edited by Y. Sawaragi, K. Inoue and H. Nakayama. XII, 445 pages. 1987.

Vol. 286: Toward Interactive and Intelligent Decision Support Systems. Volume 2. Proceedings, 1986. Edited by Y. Sawaragi, K. Inoue and H. Nakayama. XII, 450 page 1987.

Vol. 287: Dynamical Systems. Proceedings, 1985. Edite by A.B. Kurzhanski and K. Sigmund. VI, 215 pages. 198

Vol. 288: G.D. Rudebusch, The Estimation c Macroeconomic Disequilibrium Models with Regim Classification Information. VII, 128 pages. 1987.

Vol. 289: B.R. Meijboom, Planning in Decentralized Firm X, 168 pages. 1987.

Vol. 290: D.A. Carlson, A. Haurie, Infinite Horizon Opt mal Control. XI, 254 pages. 1987.

Vol. 291: N. Takahashi, Design of Adaptive Organization VI, 140 pages. 1987.

Vol. 292: I. Tchijov, L. Tomaszewicz (Eds.), Input-Outp Modeling. Proceedings, 1985. VI, 195 pages. 1987.

Vol. 293: D. Batten, J. Casti, B. Johansson (Eds.), Econom Evolution and Structural Adjustment. Proceedings, 198 VI, 382 pages.

Vol. 294: J. Jahn, W. Knabs (Eds.), Recent Advances an Historical Development of Vector Optimization. VII, 40 pages. 1987.

Vol. 295. H. Meister, The Purification Problem fc Constrained Games with Incomplete Information. X, 12 pages. 1987.

Vol. 296: A. Börsch-Supan, Econometric Analysis Discrete Choice. VIII, 211 pages. 1987.

Vol. 297: V. Fedorov, H. Läuter (Eds.), Model-Oriente Data Analysis. Proceedings, 1987. VI, 239 pages. 1988.

Vol. 298: S.H. Chew, Q. Zheng, Integral Glob Optimization. VII, 179 pages. 1988.

Vol. 299: K. Marti, Descent Directions and Efficie Solutions in Discretely Distributed Stochastic Program XIV, 178 pages. 1988.

Vol. 300: U. Derigs, Programming in Networks and Graph XI, 315 pages. 1988.

Vol. 301: J. Kacprzyk, M. Roubens (Eds.), Nor Conventional Preference Relations in Decision Makin VII, 155 pages. 1988.

Vol. 302: H.A. Eiselt, G. Pederzoli (Eds.), Advances Optimization and Control. Proceedings, 1986. VIII, 37 pages. 1988.

Vol. 303: F.X. Diebold, Empirical Modeling of Exchang Rate Dynamics. VII, 143 pages. 1988.

Vol. 304: A. Kurzhanski, K. Neumann, D. Pallaschk (Eds.), Optimization, Parallel Processing and Application Proceedings, 1987. VI, 292 pages. 1988.

Vol. 305: G.-J.C.Th. van Schijndel, Dynamic Firm an Investor Behaviour under Progressive Personal Taxatio X, 215 pages. 1988.

Vol. 306: Ch. Klein, A Static Microeconomic Model c Pure Competition. VIII, 139 pages. 1988.

Vol. 307: T.K. Dijkstra (Ed.), On Model Uncertainty an its Statistical Implications. VII, 138 pages. 1988.

Vol. 308: J.R. Daduna, A. Wren (Eds.), Computer-Aide Transit Scheduling. VIII, 339 pages. 1988.

Vol. 309: G. Ricci, K. Velupillai (Eds.), Growth Cycle and Multisectoral Economics: the Goodwin Tradition. II 126 pages. 1988.

Vol. 310: J. Kacprzyk, M. Fedrizzi (Eds.), Combining Fuzzy Imprecision with Probabilistic Uncertainty in Decision Making. IX, 399 pages. 1988.

Vol. 311: R. Färe, Fundamentals of Production Theory. IX, 163 pages. 1988.

Vol. 312: J. Krishnakumar, Estimation of Simultaneous Equation Models with Error Components Structure. X, 357 pages. 1988.

Vol. 313: W. Jammernegg, Sequential Binary Investment Decisions. VI, 156 pages. 1988.

Vol. 314: R. Tietz, W. Albers, R. Selten (Eds.), Bounded Rational Behavior in Experimental Games and Markets. VI, 368 pages. 1988.

Vol. 315: I. Orishimo, G.J.D. Hewings, P. Nijkamp (Eds), Information Technology: Social and Spatial Perspectives. Proceedings 1986. VI, 268 pages. 1988.

Vol. 316: R.L. Basmann, D.J. Slottje, K. Hayes, J.D. Johnson, D.J. Molina, The Generalized Fechner-Thurstone Direct Utility Function and Some of its Uses. VIII, 159 pages. 1988.

Vol. 317: L. Bianco, A. La Bella (Eds.), Freight Transport Planning and Logistics. Proceedings, 1987. X, 568 pages. 1988.

Vol. 318: T. Doup, Simplicial Algorithms on the Simplotope. VIII, 262 pages. 1988.

Vol. 319: D.T. Luc, Theory of Vector Optimization. VIII, 173 pages. 1989.

Vol. 320: D. van der Wijst, Financial Structure in Small Business. VII, 181 pages. 1989.

Vol. 321: M. Di Matteo, R.M. Goodwin, A. Vercelli (Eds.), Technological and Social Factors in Long Term Fluctuations. Proceedings. IX, 442 pages. 1989.

Vol. 322: T. Kollintzas (Ed.), The Rational Expectations Equilibrium Inventory Model. XI, 269 pages. 1989.

Vol. 323: M.B.M. de Koster, Capacity Oriented Analysis and Design of Production Systems. XII, 245 pages. 1989.

Vol. 324: I.M. Bomze, B.M. Pötscher, Game Theoretical Foundations of Evolutionary Stability. VI, 145 pages. 1989.

Vol. 325: P. Ferri, E. Greenberg, The Labor Market and Business Cycle Theories. X, 183 pages. 1989.

Vol. 326: Ch. Sauer, Alternative Theories of Output, Unemployment, and Inflation in Germany: 1960–1985. XIII, 206 pages. 1989.

Vol. 327: M. Tawada, Production Structure and International Trade. V, 132 pages. 1989.

Vol. 328: W. Güth, B. Kalkofen, Unique Solutions for Strategic Games. VII, 200 pages. 1989.

Vol. 329: G. Tillmann, Equity, Incentives, and Taxation. VI, 132 pages. 1989.

Vol. 330: P.M. Kort, Optimal Dynamic Investment Policies of a Value Maximizing Firm. VII, 185 pages. 1989.

Vol. 331: A. Lewandowski, A.P. Wierzbicki (Eds.), Aspiration Based Decision Support Systems. X, 400 pages. 1989.

Vol. 332: T.R. Gulledge, Jr., L.A. Litteral (Eds.), Cost Analysis Applications of Economics and Operations Research. Proceedings. VII, 422 pages. 1989.

Vol. 333: N. Dellaert, Production to Order. VII, 158 pages. 1989.

Vol. 334: H.-W. Lorenz, Nonlinear Dynamical Economics and Chaotic Motion. XI, 248 pages. 1989.

Vol. 335: A.G. Lockett, G. Islei (Eds.), Improving Decision Making in Organisations. Proceedings. IX, 606 pages. 1989.

Vol. 336: T. Puu, Nonlinear Economic Dynamics. VII, 119 pages. 1989.

Vol. 337: A. Lewandowski, I. Stanchev (Eds.), Methodology and Software for Interactive Decision Support. VIII, 309 pages. 1989.

Vol. 338: J.K. Ho, R.P. Sundarraj, DECOMP: an Implementation of Dantzig-Wolfe Decomposition for Linear Programming. VI, 206 pages.

Vol. 339: J. Terceiro Lomba, Estimation of Dynamic Econometric Models with Errors in Variables. VIII, 116 pages. 1990.

Vol. 340: T. Vasko, R. Ayres, L. Fontvieille (Eds.), Life Cycles and Long Waves. XIV, 293 pages. 1990.

Vol. 341: G.R. Uhlich, Descriptive Theories of Bargaining. IX, 165 pages. 1990.

Vol. 342: K. Okuguchi, F. Szidarovszky, The Theory of Oligopoly with Multi-Product Firms. V, 167 pages. 1990.

Vol. 343: C. Chiarella, The Elements of a Nonlinear Theory of Economic Dynamics. IX, 149 pages. 1990.

Vol. 344: K. Neumann, Stochastic Project Networks. XI, 237 pages. 1990.

Vol. 345: A. Cambini, E. Castagnoli, L. Martein, P Mazzoleni, S. Schaible (Eds.), Generalized Convexity and Fractional Programming with Economic Applications. Proceedings, 1988. VII, 361 pages. 1990.

Vol. 346: R. von Randow (Ed.), Integer Programming and Related Areas. A Classified Bibliography 1984–1987. XIII, 514 pages. 1990.

Vol. 347: D. Ríos Insua, Sensitivity Analysis in Multiobjective Decision Making. XI, 193 pages. 1990.

Vol. 348: H. Störmer, Binary Functions and their Applications. VIII, 151 pages. 1990.

Vol. 349: G.A. Pfann, Dynamic Modelling of Stochastic Demand for Manufacturing Employment. VI, 158 pages. 1990.

Vol. 350: W.-B. Zhang, Economic Dynamics. X, 232 pages. 1990.

Vol. 351: A. Lewandowski, V. Volkovich (Eds.), Multiobjective Problems of Mathematical Programming. Proceedings, 1988. VII, 315 pages. 1991.

Vol. 352: O. van Hilten, Optimal Firm Behaviour in the Context of Technological Progress and a Business Cycle. XII, 229 pages. 1991.

Vol. 353: G. Ricci (Ed.), Decision Processes in Economics. Proceedings, 1989. III, 209 pages 1991.

Vol. 354: M. Ivaldi, A Structural Analysis of Expectation Formation. XII, 230 pages. 1991.

Vol. 355: M. Salomon. Deterministic Lotsizing Models for Production Planning. VII, 158 pages. 1991.

Vol. 356: P. Korhonen, A. Lewandowski, J . Wallenius (Eds.), Multiple Criteria Decision Support. Proceedings, 1989. XII, 393 pages. 1991.

Vol. 357: P. Zörnig, Degeneracy Graphs and Simplex Cycling. XV, 194 pages. 1991.

Vol. 358: P. Knottnerus, Linear Models with Correlated Disturbances. VIII, 196 pages. 1991.

Vol. 359: E. de Jong, Exchange Rate Determination and Optimal Economic Policy Under Various Exchange Rate Regimes. VII, 270 pages. 1991.

Vol. 360: P. Stalder, Regime Translations, Spillovers and Buffer Stocks. VI, 193 pages . 1991.

Vol. 361: C. F. Daganzo, Logistics Systems Analysis. X, 321 pages. 1991.

Vol. 362: F. Gehrels, Essays In Macroeconomics of an Open Economy. VII, 183 pages. 1991.

Vol. 363: C. Puppe, Distorted Probabilities and Choice under Risk. VIII, 100 pages . 1991

Vol. 364: B. Horvath, Are Policy Variables Exogenous? XII, 162 pages. 1991.

Vol. 365: G. A. Heuer, U. Leopold-Wildburger. Balanced Silverman Games on General Discrete Sets. V, 140 pages. 1991.

Vol. 366: J. Gruber (Ed.), Econometric Decision Models. Proceedings, 1989. VIII, 636 pages. 1991.

Vol. 367: M. Grauer, D. B. Pressmar (Eds.), Parallel Computing and Mathematical Optimization. Proceedings. V, 208 pages. 1991.

Vol. 368: M. Fedrizzi, J. Kacprzyk, M. Roubens (Eds.), Interactive Fuzzy Optimization. VII, 216 pages. 1991.

Vol. 369: R. Koblo, The Visible Hand. VIII, 131 pages.1991.

Vol. 370: M. J. Beckmann, M. N. Gopalan, R. Subramanian (Eds.), Stochastic Processes and their Applications. Proceedings, 1990. XLI, 292 pages. 1991.

Vol. 371: A. Schmutzler, Flexibility and Adjustment to Information in Sequential Decision Problems. VIII, 198 pages. 1991.

Vol. 372: J. Esteban, The Social Viability of Money. X, 202 pages. 1991.

Vol. 373: A. Billot, Economic Theory of Fuzzy Equilibria. XIII, 164 pages. 1992.

Vol. 374: G. Pflug, U. Dieter (Eds.), Simulation and Optimization. Proceedings, 1990. X, 162 pages. 1992.

Vol. 375: S.-J. Chen, Ch.-L. Hwang, Fuzzy Multiple Attribute Decision Making. XII, 536 pages. 1992.

Vol. 376: K.-H. Jöckel, G. Rothe, W. Sendler (Eds.), Bootstrapping and Related Techniques. Proceedings, 1990. VIII, 247 pages. 1992.

Vol. 377: A. Villar, Operator Theorems with Applications to Distributive Problems and Equilibrium Models. XVI, 160 pages. 1992.

Vol. 378: W. Krabs, J. Zowe (Eds.), Modern Methods of Optimization. Proceedings, 1990. VIII, 348 pages. 1992.

Vol. 379: K. Marti (Ed.), Stochastic Optimization. Proceedings, 1990. VII, 182 pages. 1992.

Vol. 380: J. Odelstad, Invariance and Structural Dependence. XII, 245 pages. 1992.

Vol. 381: C. Giannini, Topics in Structural VAR Econometrics. XI, 131 pages. 1992.

Vol. 382: W. Oettli, D. Pallaschke (Eds.), Advances Optimization. Proceedings, 1991. X, 527 pages. 1992.

Vol. 383: J. Vartiainen, Capital Accumulation in Corporatist Economy. VII, 177 pages. 1992.

Vol. 384: A. Martina, Lectures on the Economic Theory Taxation. XII, 313 pages. 1992.

Vol. 385: J. Gardeazabal, M. Regúlez, The Monetary Mod of Exchange Rates and Cointegration. X, 194 pages. 199

Vol. 386: M. Desrochers, J.-M. Rousseau (Eds.), Compu ter-Aided Transit Scheduling. Proceedings, 1990. XIII, 43 pages. 1992.

Vol. 387: W. Gaertner, M. Klemisch-Ahlert, Social Choic and Bargaining Perspectives on Distributive Justice. VII 131 pages. 1992.

Vol. 388: D. Bartmann, M. J. Beckmann, Inventory Contro XV, 252 pages. 1992.

Vol. 389: B. Dutta, D. Mookherjee, T. Parthasarathy, T Raghavan, D. Ray, S. Tijs (Eds.), Game Theory an Economic Applications. Proceedings, 1990. ??, ?? pages 1992.

Vol. 390: G. Sorger, Minimum Impatience Theorem fo Recursive Economic Models. X, 162 pages. 1992.

Vol. 391: C. Keser, Experimental Duopoly Markets wit Demand Inertia. X, 150 pages. 1992.

Vol. 392: K. Frauendorfer, Stochastic Two-Stag Programming. VIII, 228 pages. 1992.

Vol. 393: B. Lucke, Price Stabilization on Worl Agricultural Markets. XI, 274 pages. 1992.

Vol. 394: Y.-J. Lai, C.-L. Hwang, Fuzzy Mathematica Programming. XIII, 301 pages. 1992.

Vol. 395: G. Haag, U. Mueller, K. G. Troitzsch (Eds.) Economic Evolution and Demographic Change. XVI, 40 pages. 1992.

Vol. 396: R. V. V. Vidal (Ed.), Applied Simulate Annealing. VIII, 358 pages. 1992.

Vol. 397: J. Wessels, A. P. Wierzbicki (Eds.), User Oriented Methodology and Techniques of Decisior Analysis and Support. Proceedings, 1991. XII, 295 pages 1993.

Vol. 398: J.-P. Urbain, Exogeneity in Error Correctior Models. XI, 189 pages. 1993.

Vol. 399: F. Gori, L. Geronazzo, M. Galeotti (Eds.). Nonlinear Dynamics in Economics and Social Sciences. Proceedings, 1991. VIII, 367 pages. 1993.

Vol. 400: H. Tanizaki, Nonlinear Filters. XII, 203 pages. 1993.

Vol. 401: K. Mosler, M. Scarsini, Stochastic Orders and Applications. V, 379 pages. 1993.

Vol. 402: A. van den Elzen, Adjustment Processes for Exchange Economies and Noncooperative Games. VII, 146 pages. 1993.

Vol. 403: G. Brennscheidt, Predictive Behavior. VI, 227 pages. 1993.

GPSR Compliance

The European Union's (EU) General Product Safety Regulation (GPSR) is a set of rules that requires consumer products to be safe and our obligations to ensure this.

If you have any concerns about our products, you can contact us on ProductSafety@springernature.com

In case Publisher is established outside the EU, the EU authorized representative is:

Springer Nature Customer Service Center GmbH
Europaplatz 3
69115 Heidelberg, Germany

Batch number: 10844675

Printed by Printforce, the Netherlands